Remote Sensing and Geographical Information Systems

Remote Sensing and Geographical Information Systems

(Basics and Applications)

Edited by

P.R. Vyas

RAWAT PUBLICATIONS

Jaipur • New Delhi • Bangalore • Hyderabad • Guwahati • Kolkata

ISBN 978-81-316-0410-6

Reprinted, 2017

The presentations, interpretations and opinions expressed by the contributors are not necessarily those of the editor.

Published by
Prem Rawat for **Rawat Publications**
Satyam Apts, Sector 3, Jawahar Nagar, Jaipur 302 004 (India)
Phone: 0141 265 1748/7006 Fax: 0141 265 1748
E-mail: info@rawatbooks.com
Website: www.rawatbooks.com

New Delhi Office
4858/24, Ansari Road, Daryaganj, New Delhi 110 002
Phone: 011 2326 3290

Also at *Bangalore, Guwahati* and *Kolkata*

Typeset by Rawat Computers, Jaipur

Printed at Nice Printing Press, New Delhi

Dedicated to
Late Professor J.L. Dashola and
Late Dr V.K. Choudhary, both of the
Department of Geography,
J.R.N. Rajasthan Vidhyapeeth, Udaipur

Contents

Part – 3
Application of Remote Sensing and GIS Techniques for Environmental and Regional Planning

Foreword

Remote Sensing and GIS Technology gained importance during the last two-and-a-half decades and is constantly being applied in planning, natural resource monitoring and management projects. Remote sensing involves study of multispectral responses of earth objects as captured by remote sensors from space. GIS is an efficient tool using input data into structured database, spatial analysis and retrieval of data. In customized environment both the technologies can be used for pre-feasibility planning of any project of the region.

The editor deserves warm congratulation for making an effort to commit to the project work carried out by the students of PG Diploma in Remote Sensing and GIS under his guidance as an internal guide, and his directorship.

Students completed their project work in various institutions of India, they have developed recent techniques for analysis of various digital data of terrain received through space technology. It is expected that this volume would be useful for planners, research scholars, students and the scientists working and studying in this field. The book provides a range of study material and applications too, that is the need of the hour.

Professor S.S. Sarangdevot
Vice Chancellor,
JRN RVU, Udaipur

Preface

The present book is divided in three parts: Part-1 is prepared as a special text for scholars focussing on the basic concepts of Remote Sensing, Digital Image Interpretation and Geographic Information System, Part-2 is the applied section dealing with the application aspects of Natural Resource Management (NRM) and Part-3 deals with the application of Remote Sensing and GIS for environment and regional planning. Remote Sensing and Geographic Information System (GIS) techniques are being used by geologists, geographers, regional planners, environmentalists in the field of geological studies, locational analysis, regional planning and natural resources planning. The basic aim of this book is to provide literature to students, research scholars, scientists and planners who are working in the field of application of Remote Sensing and GIS.

The text of the book is based on the projects undertaken by the students under my supervision and able guidance by the technical experts working in the various institutions during the internship programmes.

In Part-1, the basic concepts of Remote Sensing including EMR, Concept of Signature, Digital Image Processing and Geographic Information System and its components have been included as study material. Part-2 of the book covers application aspects of Remote Sensing and GIS in fields of land use and land cover, groundwater, natural resource management, etc.

Dr Jagdish Soni and Dr R.P. Naraniwal presented a very valuable article entitled 'Land Use/Land Cover Change Detection

Studies Using Remote Sensing and GIS Techniques: The Case of Dhari Taluka, Gujarat'. Another article was presented by Maya Poonia and Dr Sunita Singh entitled 'Groundwater Prospect Zoning by Remote Sensing and GIS Techniques in Dungarpur Block'. One of the important aspects of application of Remote Sensing is discussed by Akshay Manohar and Dr Yuv Raj Singh Rathore in their article entitled 'Natural Resources Analysis by Using Remote Sensing and GIS Techniques: A Case Study of Papum Pare District, Arunachal Pradesh'. Reena Devi and Dr L.R. Patel discussed digitization of cadastral maps and their intergration with high resolution satellite data for land use and land cover mapping in their research paper, while Neeraj Chahar worked on palaeo drainage mapping using remote sensing techniques in the area of Jhunjhunu district of Rajasthan. Another important work was carried out by Veena Charan, Amarjeet and Chanchal Choudhary entitled 'Groundwater Prospecting in Suk Tawa Sub-Watershed Using Remote Sensing and GIS Techniques'. Somendra K. Sharma and Dr Pankaj Rawal submitted a very valuable paper entitled 'Prospecting Groundwater Resources by Remote Sensing and GIS Techniques: Study of Sagwara Block, Dungarpur District'.

Current issues of environmental and regional planning have been included in Part-3 of the book. All these articles are based on project work carried out by the research scholars. They completed their project work at various prestigious departments of space technology under internship programme.

The first paper of this section is prepared by Professor P.R. Vyas and Dr R.P. Sharma for Micro Level Regional Planning entitled 'Application of Remote Sensing and GIS in Land Use and Land Cover Analysis'. Another research article is prepared by Rahul Tiwari and Pradeep Purohit entitled 'Land Use Planning of Vaniyambadi Block, Vellore District, Tamil Nadu'. Another important aspect is discussed in research paper 'Application of Remote Sensing and GIS Technique for Land Use Planning in Irrigation Project Area: A Case Study of Upper Betwa Basin of Madhya Pradesh' presented by Professor P.R. Vyas and Shankar Bagale. P.R. Vyas, Ashwini Kumar Vaishnav and J.P. Sharma also presented a research paper entitled 'Land Use/Land Cover Analysis of Teetriya Gram Panchayat Using GIS and Remote Sensing Techniques'. Finally Professor P.R. Vyas and Deepak Kumar

Ahuja have contributed a paper entitled 'Application of Remote Sensing and GIS Techniques for Land Use and Land Cover Analysis and Change Detection Study of Udaipur City'.

We express our sincere thanks to the scientists who could spare their valuable time to guide our scholars during internship completed at their institute specially Dr. G. Sreenivasan, Scientist, PRSSC, DOS-ISRO, Nagpur, Dr. K.R. Khichar, Remote Sensing Division, B.M. Birla Research and Technology Centre, Jaipur, Dr. Vijay Singh, Senior Scientist, BISAG, Gandhinagar, Dr. B.V. Rama Kumar, General Manager, Remote Sensing Instruments, Hyderabad, Dr. V.S. Arya, Senior Scientist, SG Soil Survey and Land Evaluation, CCS HAU, Hisar (Haryana), Dr. M.P. Punia, Head, Remote Sensing Division, B.M. Birla Science and Technology Centre, Jaipur, Sudheer Bhatnagar, Directorate, Mines and Geology Department, Udaipur, Dr. V. Madhav Rao, Professor and Head, CGARD, NIRD, Hyderabad, Dr. V.V. Rao, Scientist, SF, NRSA, Hyderabad, Rakesh Paliwal, Scientist, RRSC, ISRO, Department of Space, Government of India, Jodhpur, Pradeep Bhatnagar, Superintending Geologist, RS, Mines and Geology Department, Udaipur.

Finally, we are also thankful to faculty members of the department Dr. Sunita Singh, Head and Associate Professor for the co-operation extended during editing work, Dr. R.P. Naraniwal, Associate Professor and Dr. L.R. Patel, Associate Professor for their valuable suggestions, Dr. Pankaj Rawal, Assistant Professor, Yuvraj Singh, Assistant Professor and Dr. J.P. Sharma (former faculty) as Assistant Professor for their help in course work.

It is expected that the present volume will serve the scholars working in the field of remote sensing and GIS. As we know that the Remote Sensing Technology provides not only numerous scientific applications and benefits that can be used efficiently but also helps in the day-to-day decision-making processes in an ever-increasing range of applications. The present book intends to provide a basic knowledge as well as application techniques towards making our planet earth's resources more sustainable.

P.R. Vyas

Acronyms

ArcEditor This is the mid-level software suite designed for advanced editing of spatial data published in the proprietary Esri format. It is part of the ArcGIS product. It provides tools for the creation of map and spatial data used in Geospatial Information Systems.

ArcInfo A full-featured geographic information system produced by ESRI. The first version of ARC/INFO was launched in 1982 on minicomputers, as ESRI claims, the very first modern GIS. The name refers to its architecture as a Geographic Information System composed of (a) geographic input, processing, and output tools ("ARC") with (b) a complementary, but separate database ("INFO").

ArcView It is the entry level licensing level of ArcGIS Desktop, a geographic information system software product produced by Esri. ArcView is now referred to as ArcGIS for Desktop Basic.

AutoCAD A software application for both 2D and 3D computer-aided design (CAD) and drafting.

AWiFS Advanced Wide Field Sensor (AWiFS) operating in three spectral bands in VNIR and one band in SWIR with 56 metre spatial resolution.

DN Digital number, is the discrete representation of an analog value sampled by an analog-to-digital converter.

DPI	Dots per inch is a measure of spatial printing or video dot density, in particular the number of individual dots that can be placed in a line within the span of 1 inch (2.54 cm).
DTM	Digital Terrain Modelling
ERDAS	ERDAS has been supplying image processing software since 1978. Their flagship desktop package, ERDAS IMAGINE, led the way in image processing and raster data handling. It enables the user to extract geospatial information from the raw satellite or aerial imagery.
ESRI	Esri is an international supplier of Geographic Information System (GIS) software and geo-database management applications. The company was founded as Environmental Systems Research Institute (ESRI) in 1969 as a land-use consulting firm.
ETM	Enhanced Thematic Mapper, an instrument on board the Landsat 7.
FCC	False colour composite
GPS	Global Positioning System is a space-based satellite navigation system that provides location and time information in all weather conditions, anywhere on or near the Earth where there is an unobstructed line of sight to four or more GPS satellites.
ICRISAT	International Crops Research Institute for the Semi-Arid-Tropics is a non-profit agricultural research organization headquartered in Patancheru (Hyderabad Andhra Pradesh, India) with several regional centres: Niamey (Niger), Nairobi (Kenya) and research stations Bamako (Mali), Bulawayo (Zimbabwe).
IRS-1B	It was the first remote sensing mission undertaken by ISRO using the imagery generated by remote sensing technology in land based applications.
IRS-1C	This was the fourth remote sensing Indian satellite built, and designed by Indian Space Research Organization (ISRO). The primary objective of IRS-1C was to provide systematic and repetitive acquisition of data of the earth's surface under nearly constant illumination conditions.
IRS-1D	It is the seventh satellite in Indian Remote Sensing satellite series of earth observation satellites, built,

launched and maintained by Indian Space Research Organisation. The satellite has similar capabilities as IRS-1C satellite with some improvements added for better imagery, particularly in thematic mapping.

IRS-P6 IRS-P6 generally known as RESOURCESAT-1 is an advanced remote sensing satellite built by Indian Space Research Organization. This tenth satellite of ISRO in IRS series, is intended to not only continue the remote sensing data services provided by IRS-1C and IRS-1D, both of which have far outlived their designed mission lives, but also vastly enhance the data quality.

ISRO Indian Space Research Organisation is amongst the six largest government space agencies in the world, along with USA's NASA, Russia's RKA, Europe's ESA, China's CNSA and Japan's JAXA. Its primary objective is to advance space technology and use its applications for national benefit. ISRO is under the administrative control of the Department of Space, Government of India.

Landsat (TM) Landsat (Thematic Mapper) is a multispectral scanning radiometer that was carried on board Landsats 4 and 5. The TM sensors have provided nearly continuous coverage from July 1982 to present. The Landsat Enhanced Thematic Mapper (ETM) was introduced with Landsat 7.

LCC Projection A Lambert Conformal Conic projection is a conic map projection, which is often used for aeronautical charts. In essence, the projection seats a cone over the sphere of the Earth and projects conformably onto the cone.

LISS-4 A high resolution Liear Imaging Self Scanner operating in three spectral bands in the Visible and Near Infrared Region (VNIR) with 5.8 metre spatial resolution and steerable up to 26 degrees across track to obtain stereo-scopic imagery and achieve five day revisit capability.

Lpd Litres per day (water treatment system output).

Lpm refers to Litre per minute, a volumetric flow rate.

LULC Land Use and Land Cover classification

NASA National Aeronautics and Space Administration is the agency of the United States government that is

	responsible for the nation's civilian space programme and for aeronautics and aerospace research.
NDVI	Normalized Difference Vegetation Index is a simple graphical indicator that can be used to analyse remote sensing measurements, typically but not necessarily from a space platform, and assess whether the target being observed contains live green vegetation or not.
NOAA	National Oceanic and Atmospheric Administration is a scientific agency within the United States Department of Commerce focused on the conditions of the oceans and the atmosphere.
NRSA	refers to the present National Remote Sensing Centre (NRSC). A central organization of the government of India responsible for managing data from imaging satellites.
SOI	Survey of India is India's central engineering agency in charge of mapping and surveying.
SWIR	Short Wave Infrared
USGS	United States Geological Survey is a scientific agency of the United States government.
UTM	The Universal Transverse Mercator geographic coordinate system uses a 2-dimensional Cartesian coordinate system to give locations on the surface of the earth.
VNIR	Visible and Near Infrared Region
WGS 1984	The World Geodetic System is a standard for use in cartography, geodesy, and navigation. WGS 84 is the reference coordinate system used by the Global Positioning System.
WUI	Water Utilization Index is defined as the ratio of the crop area irrigated in an irrigation command and the water delivered to that irrigation command.

Part – 1

Basics of Remote Sensing and GIS

1

Principles of Remote Sensing

Shankar Bagale and Amarjeet

Introduction

We may call "Remote Sensing is a science as well as an art of obtaining information about an object, area or phenomenon through an analysis of the data acquired by a device which is not in physical contact with the object, area or phenomenon under investigation". Such sensing may be achieved in the absence of any matter in the intervening space between the object and the observer. The information carrier or communication link is electromagnetic energy. Remote sensing data basically consists of wavelength intensity information acquired by collecting the electromagnetic radiation leaving the object at specific wavelength and measuring its intensity. Hence, currently the term "Remote Sensing" is used more commonly to denote, "identification of earth features by detecting the characteristic of electromagnetic radiation that is reflected or emitted by the earth surface".

Remote sensing of earth's features comprises measuring and recording of electromagnetic energy reflected from or emitted by the planet's surface and atmosphere from a vantage point above the surface, and relation of such measurement to the nature and distribution of surface materials and atmospheric conditions.

The electromagnetic energy emitted or reflected from an object or geographic area is used as an image for the actual property under investigation. The electromagnetic energy measurements must be turned into information using visual and/or digital image processing techniques. The majority of remote sensing instruments record electromagnetic radiation (EMR) that travels at a velocity of 3×10^8 m/sec from the source, directly through the vacuum of space or indirectly by reflection or re-radiation to the sensor. The EMR represents an extremely efficient high-speed communication link between the sensor and the remotely located phenomenon. However, the majority of remotely sensed data collected for earth resource applications are the result of sensors that record electromagnetic energy.

Major Components of Remote Sensing Technology

The following are major components of remote sensing system:

(i) *Energy Source*
   (a) Passive System: sun irradiance from earth's materials;
   (b) Active System: irradiance from artificially generated energy sources such as radar.

(ii) *Platform*: Vehicle to carry the sensor (truck, aircraft, space shuttle, satellite, etc.).

(iii) *Sensor*: Device to detect electromagnetic radiation (camera, scanner, etc.).

(iv) *Detector*: Device for collection of signal data (photographic, digital, etc.).

(v) *Processing*: Handling of signal data (photographic, digital, etc.).

(vi) *Institutionalization*: Organization for execution at all stages of remote-sensing technology: international and national organizations, centres, universities, etc.

Stages in Remote Sensing

Remote sensing largely concerns with the measurement of electromagnetic energy, which is reflected and scattered or emitted by the objects. Different objects return different sorts and amounts of energy. The detection of these differences enables identification of the features on the ground. The remote sensing system consists of a sensor mounted on a platform (viz., satellite, rocket, aircraft, etc.) to collect the radiation from the ground. The information received by the sensor is suitably manipulated and sent back to the earth. The data are processed on the ground to produce either photographs or digital data. The photographs or digital data are interpreted visually or digitally to produce thematic maps and other reference sources.

Definition

Remote sensing is the science and art of collecting information about an object, area, or phenomenon through the analysis of data acquired by a device that is not in contact with the object, area, or phenomenon under investigation (Lillesand and Kiefer, 1994).

Stages of Remote Sensing

The different stages that are involved in remote sensing are as stated below:

(i) Origin of electromagnetic energy (sun, transmitter carried by the sensor).

(ii) Transmission of energy from the sources to the surface of the earth and its interaction with the intervening atmosphere.

(iii) Interaction of energy with surface of the earth, reflection/absorption/transmission/or self-emission.

Figure 1.1

Concept of Remote Sensing

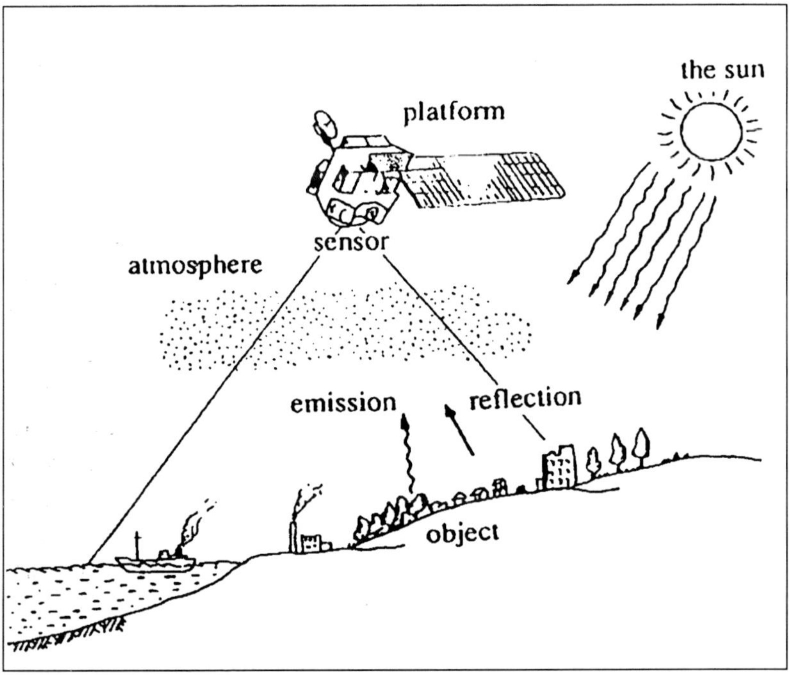

(iv) Transmission of the reflected/emitted energy to the remote sensor placed on a suitable platform (aircraft, space shuttle, satellite, etc.).

(v) Detection of the energy by the sensor camera, scanner, etc., converting it into photographic image or equivalent electrical output.

(vi) Transmission/recording of the sensor output.

(vii) Preprocessing of the data for generation of the data products.

(viii) Collection of ground truth and other collateral information.

(ix) Data processing and interpretation.

The concept of remote sensing is shown in Figure 1.1.

Electromagnetic Energy (EMR)

In remote sensing, electromagnetic energy is referred similarly with light: velocity of light (3×10^8 m/s) in a harmonic wave (equally and repetitively spaced in time) and has two force fields, electric and magnetic to each other (as shown in Figure 1.2). From basic physics, velocity of light (c) = frequency (f) × wavelength (λ).

Figure 1.2

Electromagnetic Wave Pattern

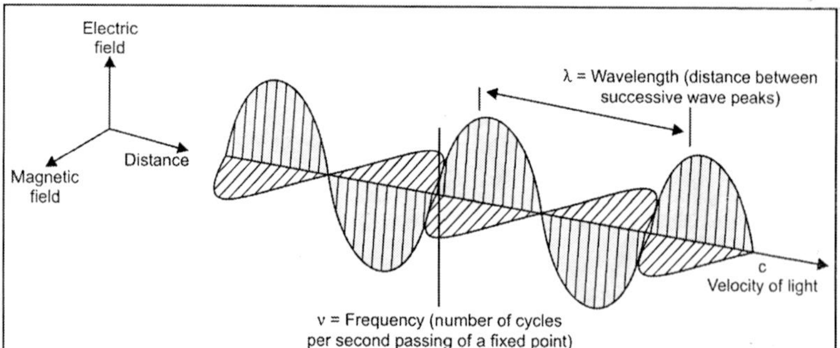

An electromagnetic wave components include a sinusoidal electric wave (E) and a similar magnetic wave (M) at right angles, both using perpendicular to the direction of propagation.

The electromagnetic radiation (EMR) is composed by discrete units called photons or quanta. The energy of a quantum is given as

$$E = hf$$

where

E = Energy of a quantum in Joules (J)

h = Planck's constant = 6.626×10^{-34} J-sec of energy of EMR

f = frequency of wavelength

Electromagnetic Spectrum

When all of the possible forms of radiation are classified and arranged according to wavelength or frequency, the result is the electromagnetic spectrum. The electromagnetic spectrum includes types of radiation that range from extremely low energy, long wavelength, low frequency energy like radio energy to extremely high energy, short wavelength, high frequency energy types such as X-ray and Gamma-ray radiation. Wavelength regions of electromagnetic radiation have different names ranging from Gamma-ray, X-ray, Ultraviolet (UV), Visible light, Infrared (IR) to Radio Wave, These regions are shown in Figure 1.3. Some of the basic properties are enlisted in the Table 1.1.

Table 1.1

Basic Properties of Electromagnetic Spectrum

S. No.	*Wavelength*	*Nominal Spectral Location*	*Properties*
1	0.40–0.50	Blue	For water body penetration, and coastal water analysis. Also useful for soil/vegetation discrimination, identifying forest types, and cultural feature identification.
2	0.50–0.60	Green	Measuring green reflectance peak of vegetation for vegetation discrimination and vigour assessment. Also for cultural feature identification.
3	0.60–0.69	Red	To sense in a chlorophyll absorption region aiding in plant species differentiation. For cultural feature identification.
4	0.75–0.90	Near infrared	For determining vegetation types, vigour, and biomass content. For delineating water bodies, and for soil moisture discrimination.
5	1.55–1.75	Mid-infrared	Indicative of vegetation moisture content and soil moisture discriminations, and thermal mapping applications.
6	10.5–12.5	Thermal infrared	Vegetation stress analysis, soil moisture discrimination, and thermal mapping applications.
7	2.08–2.35	Mid-infrared	For discrimination of mineral and rock types. Also sensitive to vegetation moisture content.

EMR Interaction with Earth Surface

When electromagnetic energy is incident on any given earth surface feature, three fundamental energy interactions with the feature are possible. Various fractions of the energy incident on the element are reflected, absorbed and transmitted. Applying the Law of Conservation of Energy:

$$T + \quad + \rho = 1$$

- Transmittance (T): Some fraction (up to 100%) of the radiation penetrates into certain surface materials such as water and if the material is transparent and thin in one dimension, normally passes through, generally with some diminution.
- Absorption ( ): Some radiation is absorbed through electron or molecular reactions within the medium; a portion of this energy is then re-emitted, usually at longer wavelengths, and some of it remains and heats the target;
- Reflectance (ρ): Some radiation (commonly 100%) reflects (moves away from the target) and scatters away from the target at various angles, depending on the surface roughness and the angle of incidence of the rays.

Figure 1.3

Electromagnetic Spectrum

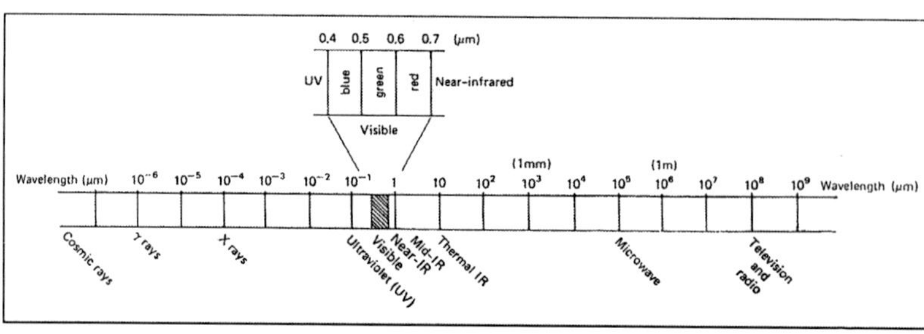

Spectral Reflectance of Vegetation, Soil, Water

Figure 1.4 shows typical spectral reflectance curves for three basic earth features, i.e., healthy green vegetation, dry bare soil and clear water. The line in the figure represents average reflectance curves compiled by measuring a large sample of features. In general, the

Figure 1.4

Wavelengths (μm)

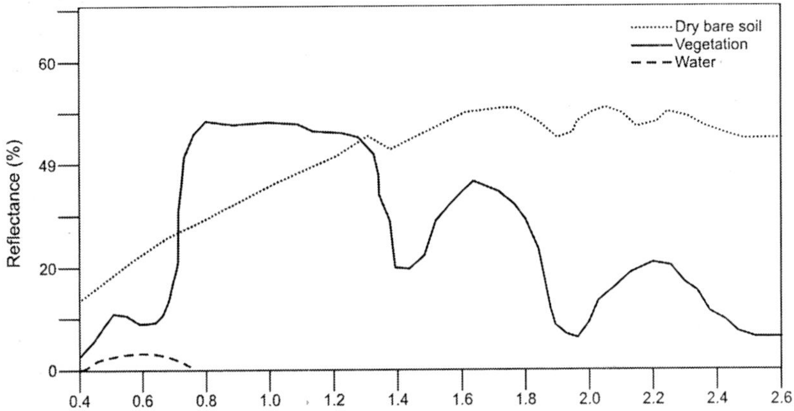

configuration of these curves is an indicator of the type and condition of the features to which they apply. This spectral reflectance concept is used for interpretation of satellite data.

(i) Vegetation – in *Green band*: The reflectance in green band, i.e., 0.45 to 0.65 μm region is more as compared to red and blue region and is also more in the region from 0.7 μm to 1.3 μm of the Near-infrared. The pigments in the plant leaves, Chlorophyll is strong in the absorb energy level between the wavelength 0.45 μm and 0.65 μm. Therefore, our eyes perceive healthy vegetation as green because of high absorption of blue and red energy by plant leaves and very high reflectance in green energy; if that decreases in the chlorophyll it results in absorption of blue and red bands. Often the red reflectance increases to the point that we see the plant turn yellow (combination of green and red).

In *Near-Infrared (NIR band):* As we go from visible region to infrared portion of the spectrum at about 0.7 μm the reflectance of healthy vegetation increases dramatically. In the range 0.7 to 1.3 μm (under NIR band), a plant leaf reflects about 50 per cent of the energy incident upon it, and remaining is transmitted. The reflectance in 0.7 μm to 1.3 μm range results primarily from the internal structure of the plant leaves, because this structure is highly variable between plant species; reflectance measurements in this range often permit us to discriminate between species, even if they look the same in the visible wavelength. This is the reason for using NIR, red bands for calculating Normalized Difference Vegetation Index (NDVI). And also the same theory is used in this study to discriminate in paddy crop as early, mid and late using in 0.7 μm to 1.3 μm region of EMR spectrum using IRS-1C, LISS-III data.

In *Mid-Infrared*: Beyond 1.3 μm energy incident upon vegetation is essentially absorbed or reflected, with little to no transmittance of energy. Dips in reflectance occur at 1.4, 1.9, 2.7 μm because water in the leaf absorbs these wavelengths strongly. Accordingly, wavelengths in these spectral regions are referred to as water absorption bands. Reflectance peaks occurred at above 1.6 and 2.2 μm between the absorption bands. Throughout the range beyond 1.3 μm, leaf reflectance is approximately inversely related to the total water percentage in the leaf. This total is a function of the moisture content and the thickness of a leaf.

(ii) Soil: Soil curve in Figure 1.4 shows considerably less variation, the main factors that influence the soil reflectance act over less specific spectral bands; some of the factors affecting soil reflectance are moisture content, soil texture, surface roughness, and presence of iron oxide and organic matter content. This factor is complex, variable and interrelated. For example, the presence of moisture in the soil will decrease its reflectance. As with vegetation this effect is greatest in the water absorption bands at about 1.4, 1.9, and 2.7 μm. Soil moisture content is strongly related to the soil texture – coarse, sandy soils are well drained, resulting in low moisture content and have relatively high reflectance; poorly drained fine textured soil will generally have lower reflectance. In the absence of water, however, soil itself will exhibit the reverse tendency; coarse textured soil will appear darker than fine textured soil. Thus, reflectance properties of a soil are consistent only within particular ranges of conditions. Two other factors that reduce the soil reflectance are surface roughness and content of organic matter.

(iii) Water: Water absorbs the energy in infrared wavelength of the various water features (such as lakes and streams) or water contained in the vegetation or soil. Locating and delineating water bodies with remote sensing data is done most easily in reflected infrared wavelengths because of these absorption properties. However, various conditions of water bodies manifest themselves in visible wavelength, e.g., the reflectance of a water body span from an interaction with the water surface with material suspended in the water, or with the bottom of the water body. Even with deep water, where bottom effects are negligible, the reflectance properties of a water body are not only a function of the water per se, but also effect by diluted material in the water.

Clear water absorbs relatively little energy since it has a wavelength of less than about 0.6 μm. High transmittance typifies these wavelengths with a maximum in blue and green portion of the spectrum; however, as turbidity of water changes transmittance (because of the presence of the organic or inorganic materials), reflectance changes dramatically. For example, water containing large quantities of suspended sediments resulting from soil erosion, normally has much higher visible reflectance than other clear waters in the same geographical area. Likewise, the reflectance of water changes with the chlorophyll concentration involved. Increases in chlorophyll concentration tend to decrease water reflectance in blue wavelengths and increase it in the green wavelength. These changes have been used to monitor the presence and estimate the concentration of algae via remote sensing data.

Remote Sensing Satellites and Platforms

Resolution

In general, resolution is defined as the ability of an entire remote sensing system, including lens antennae, display, exposure, processing, and other factors, to render a sharply defined image. While selecting any satellite data, the resolution point should be principally considered. Resolution in remote sensing is of different types.

(i) *Spectral Resolution* of a remote sensing instrument (sensor) is determined by the bandwidths of the electromagnetic radiation of the channels used. High spectral resolution, thus, is achieved by narrow bandwidths collectively, and is likely to provide a more accurate spectral signature for discrete objects than broad bandwidth.

(ii) *Radiometric Resolution* is determined by the number of discrete levels into which signals may be divided.

(iii) *Spatial Resolution* in terms of the geometric properties of the imaging system, is usually described as the instantaneous field of view (IFOV). The IFOV is defined as the maximum angle of view in which a sensor can effectively detect electromagnetic energy.

(iv) *Temporal Resolution* is related to the repetitive coverage of the ground by the remote-sensing system. The temporal resolution of Land Sat 4/5 is sixteen days.

Platforms and Sensors

IRS 1C

IRS 1C, the fourth in the IRS series was launched from Baikanur, Cosmodrome on May 19, 1995. It operates in a near polar, sun-synchronous orbit at an altitude of 817 km. Its local equatorial crossing time is 10:30 AM in the descending node. The satellite payload consists of three sensors, namely Panchromatic camera (PAN), Linear Imaging and Self-scanning Sensor (LISS-III) and Wide Field Sensor (WiFS).

LISS-III sensor has suitable features for the current irrigation assessment and crop studies. LISS-III camera provides multi-spectral data in 4 bands. The spatial resolution for visible (two bands) and near-infrared (one band) is 23.5 m with a ground swath of 141 km. The fourth band (short wave infrared band) has a spatial resolution of 70.5 m with a ground swath of 148 km. The revisit of LISS-III is 24 days. Characteristic of LISS-III (IRS-1C) given in Table 1.2.

Table 1.2

Characteristics of LISS-III (IRS-1C, 1D)

	IRS 1C	*IRS 1D*
Resolution	23.5 m (Visible & Nr. IR region) 70.5 m (Shortwave IR region)	21.2–23.5 m (Visible and Nr. IR region) 63.6–70.5 m (Shortwave IR region)
Swath	141 km. (Visible & Nr. IR region) 148 km. (Shortwave IR region)	127–141 km. (Visible & Nr. IR region) 133–148 km. (Shortwave IR region)
Revisit	24 days	25 days
Spectral Bands	0.52–0.59 microns (B2) 0.62–0.68 microns (B3) 0.77–0.86 microns (B4) 1.55–1.70 microns (B5)	0.52–0.59 microns (B2) 0.62–0.68 microns (B3) 0.77–0.86 microns (B4) 1.55–1.70 microns (Bt)
Quantization	7 bits	7 bits

IRS-1D

IRS-1D, fifth in the IRS series was launched from Sriharikota, India on September 29, 1996 using Polar Satellite Launch Vehicle, PSLV-C1 It carries payloads similar to its predecessor, IRS-1C. Like IRS-1C, IRS-1D has LISS-III, PAN, WiFS sensors on board.

IRS-P6

Resourcesat-1 (IRS-P6) is especially designed for integrated land and water management and agricultural applications and also envisaged as the continuity mission to IRS-1C/1D, with enhanced capabilities both in the payload and the platform, to meet the increasing demands of the user community. The satellite was launched by the indigenously built Polar Satellite Launch Vehicle on October 17, 2003. The orbital parameters of IRS-P6 are the same as that of IRS-1C. The payload system of IRS-P6 consists of three solid-state cameras:

(i) A high resolution multi-spectral sensor – LISS-IV
(ii) A medium resolution multi-spectral sensor – LISS-III
(iii) An Advanced Wide Field Sensor – AWiFS

The LISS-IV camera is a multi-spectral high resolution camera with a spatial resolution of 5.8m at nadir. The sensor consists of three linear odd-even pairs of charge coupled device (CCD) arrays, each with 12,000 pixels. Also the placement of the three CCDs in the focal plane is such that their imaging strips on the ground are separated by 14.25 km in the along-track direction. The LISS-IV camera has the additional feature of off-nadir viewing capability by tilting the camera by +/- 26 deg. Thus, it can provide a revisit of 5 days for any given ground area. This camera can be operated in two modes, Mono and Multi-spectral.

In Mono mode a corresponding swath of 70 km can be transmitted. Normally, Band-3 data are transmitted in this mode. Multi-spectral mode has a corresponding 23.9 km swath. This multi-spectral data are collected in three spectral bands as mentioned in Table 1.3.

The LISS-III is a multi-spectral camera operating in four spectral bands, three in the visible and near infra-red and one in short wavelength infrared (SWIR) region, as in the case of IRS-1C/1D. The new feature in LISS-III camera is the SWIR band (1.55 to 1.7 microns), which provides data with a spatial resolution of 23.5 m unlike IRS-1C/1D (the spatial resolution is 70 m).

Table 1.3

Characteristics of LISS-IV (IRS-P6)

IGFOV* (across track)	5.8 m
Ground sampling distance	5.8 m
Spectral Bands	0.52–0.59 microns (Green B2) 0.62–0.68 microns (Red B3) 0.76–0.86 microns (NIR B4)
Swath	23.9 km (multi-spectral mode) 70.0 km (mono mode)
Quantization	10 bits Selected 7 bits will be transmitted by the data handling system

* Instantaneous Geometric Field of View

Table 1.4

Characteristics of LISS-III (IRS-P6)

Resolution	23.5 m
Swath	141 km
Spectral Bands	0.52–0.59 (Green B2) 0.62–0.68 (Red B3) 0.77–0.86 (Near Infrared B4) 1.55–1.70 (Mid Infrared B5)
Quantization	7 bits and SWIR bands has 10 bit

Selection of Satellite Image (Image Browsing)

Before browsing and selecting a satellite image, first decide the objective of the project. Analyse the requirement, then decide and go for image browsing.

- Locate the study area for which image is required.
- Find out relevant things mentioned below about the study area:
  - Latitude and longitude of study area.
  - Major city name nearby the study area.
  - Satellite Path No.
  - Date of acquisition.
  - Toposheet No.
- Above are the input to image browsing with proper login and password.

- Select the satellite and sensor with mentioned cloud free data.
- Mention the start date and end date as per the requirement.
- Plot the available image and examine it for the required study area.
- Place order for above searched image.

Note: Select that date of image which fulfils the objective of the project. Selective image browsing websites are listed below.

Image Browsing Websites

S. No.	*Satellite/Organization*	*Web address*	*Data*
1	India, NRSA	www.nrsa.gov.in	IRS
2	Digital Globe	www.archivetool3.digitalglobe.com	Quick Bird
3	Space Imagine	ww.carterraonline.spaceimaging.com/	IKONOS
4	US	http://edcsns17.cr.usgs.gov/Earth Explorer/http://glovis.usgs.gov/	Landsat 7
5	Russia	http://sirius.spotimage.fr/cgi	Spot

References

Aronoff, S. (1989), *Geographic Information Systems: A Management Perspective,* WDL Publications, Ottawa, Canada.

Burrough, P.A. (1986), *Principles of Geographical Information Systems for Land Resources Assessment,* Clarendon Press, Oxford.

Curran, P.J. (1988), *Principles of Remote Sensing,* Longman, London.

Jensen, J.R. (2004), *Remote Sensing of the Environment: An Earth Resource Perspective,* University of South Carolina, Columbia, USA.

Joseph, G. (2004), *Fundamentals of Remote Sensing,* Cambridge University Press, Cambridge, UK.

NRSA (1995, 1999), *IRS Handbook,* National Remote Sensing Agency, Hyderabad.

Reddy, M. Angi (2001), *Remote Sensing and Geographic Information System,* B.S. Publications, Hyderabad.

Rees, W. (1990), *The Physical Principles of Remote Sensing,* Cambridge University Press, Cambridge, UK.

Reeves, R.G. (1970), *Mannual of Remote Sensing,* American Society of Photogrammetry, Falls Church, Virginia.

Sabins, F.F. (1987), *Remote Sensing: Principles and Interpretation,* 2nd Edition, W.H. Freeman & Co., New York.

2

Geographic Information System

Akshay Manohar and Rahul Tiwari

In the past twenty-five years, a group of scientists have been in the process of developing automated tools for efficient storage, analysis and presentation of geographic data. These efforts have apparently been the result of increasing demands by users for the data and information of a spatial nature. This rapidly evolving technology has come to be known as, Geographic Information Systems (GIS). The applications of this technology are vast and cut across virtually all professions. GIS is a tool that facilitates generation of information required for a particular need of discipline. Input data to the systems is geo-referenced in spatial as well as tabular form. The system provides efficient tools for inputting data into database, retrieval of selected data items for further processing and software modules which can analyse/manipulate the retrieved data in order to generate desired information on specific form.

Basically, GIS is a system for storing, retrieving and analysing geographically referenced data sets. In GIS, spatial (e.g., maps, satellite imagery) and non-spatial (e.g., census data, field data) sources can be integrated and a set of spatially registered data layers can be analysed independently or in combination with a number of layers, as is demonstrated in Figure 2.1.

Figure 2.1

Thematic Information in GIS

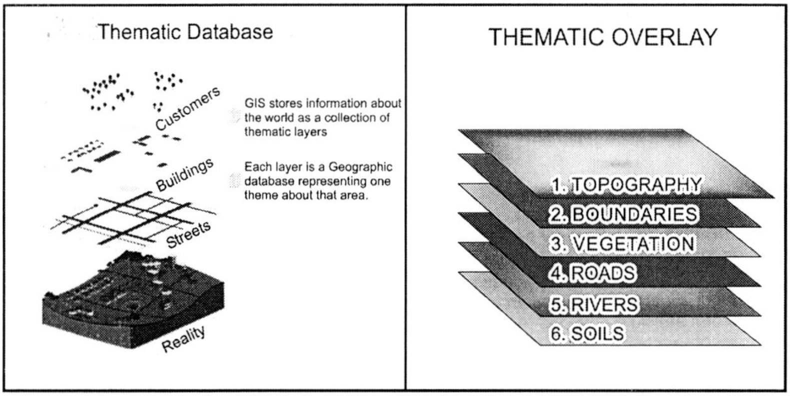

Thus, one can stitch together bundles of data from a wide variety of sources and manipulate them to get quantitative information for resources management and planning at different levels. GIS consists of such activities as capturing of data, archival, new survey data, integration of data sets, design of efficient data structure, database, user interfaces and analysing integrated data sets. The ability of GIS to identify new relationship from the existing database elements and to present the results of the analysis in cartographic form makes it exceedingly important from the planning point of view, as most macro- and micro-level planning tasks ideally require geographically referenced information.

The emphasis in information processing in the 1960s was on hardware, in the 1970s on software, in 1980s it was on database, and in 1990s computer cartography and Spatial Decision Support Systems (Artificial Neural Networks) emerged as superior tools in accessing and interpreting the GIS contents.

Definition

GIS can be understood by the help of various definitions given under:

- A geographic information system (GIS) is a computer-based tool for mapping and analysing things that exist and events that happen on Earth.

- Burrough in 1986 defined GIS as, "Set of tools for collecting, storing, retrieving at will, transforming and displaying spatial data from the real world for a particular set of purposes".
- Arnoff in 1989 defined GIS as, "a computer based system that provides four sets of capabilities to handle geo-referenced data as data input, data management (data storage and retrieval), manipulation and analysis, data output".

Hence, GIS is looked upon as a tool to assist in decision-making and management of attributes that needs to be analysed spatially.

Components of GIS

(i) *Hardware:* It consists of the computer system on which the GIS software will run. The choice of hardware system, range from 300 MHz Personal Computers to Super Computers having capability in Tera FLOPS. The computer forms the backbone of the GIS hardware, which gets its input through the scanner or a digitizer board. Scanner converts a picture into a digital image for further processing. The output of the scanner can be stored in many formats, e.g., TIFF, BMP, JPG, etc. A digitizer board is a flat board used for vectorization of a given map object. Printers and plotters are the most common output devices for a GIS hardware set-up.

(ii) *Software:* GIS software provides the functions and tools needed to store, analyse and display geographic information. GIS softwares in use are MapInfo, ARC/Info, AutoCAD Map, etc. The software available can be said to be application specific. When the low cost GIS work is to be carried out, desktop MapInfo is the suitable option. It is easy to use and supports many GIS features. If the user intends to carry out extensive analysis on GIS, ARC/Info is the preferred option. For the people using AutoCAD and willing to step into GIS, AutoCAD Map is a good option. Other packages available are Intergraph, IDRISI, ENVI, Arc-GIS, Micro-station and Small World.

(iii) *People/Users:* GIS users range from technical specialists who design and maintain the system to those who use it to help them perform their everyday work. The people who use GIS can be broadly classified into two classes. The CAD/GIS operator, whose work is to vectorize the map objects? The use of this vectorized data to perform query, analysis or any other work is the responsibility of a GIS engineer/user.

(iv) *Data:* Geographic data and related tabular data can be collected in-house or purchased from a commercial data provider. The digital map forms the basic data input for GIS. Tabular data related to the map objects can also be attached to the digital data. A GIS will integrate spatial data with other data resources and can even use a Database Management System (DBMS), used by most organization to maintain their data and to manage spatial data. It can be further described and divided into two, viz.:

   (a) *Spatial information*: Represents geographic features (location and shapes) associated with the real world locations and their relationships with other features.

   (b) *Non-spatial information*: It is descriptive or tabular information about the characteristics of the feature.

Method: And above all, a successful GIS operates according to a well-designed plan and business rules, which are the models and operating practices unique to each organization. There are various techniques used for map creation and further usage for any project. The map creation can either be automated raster to vector creator or it can be manually vectorized using the scanned images. The source of these digital maps can either be map prepared by any survey agency or by satellite imagery.

Functions of GIS

The functions of GIS can be well described under the following heads:

(i) Topologic Data Elements: The elements in a spatial database are representative of real world entities with associated attributes. The power of GIS comes from its ability to look at entities in their geographical context and examine relationships between entities. The different types of spatial elements are shown in Figure 2.2 and can be described as follows:

   (a) *Point Feature*: It is the simplest type of spatial element. Coordinates of each point can be stored as two additional attributes. Information on a set of points can be viewed as an extended attribute table. Each point is independent of every other point, represented as a separate row in the database model.

   (b) *Line Feature:* It is composed of vertices, nodes and arcs. It includes a set of ordered coordinates that when connected represent a linear feature. Examples of such data elements are infrastructure networks, transport network, drainage network, etc.

(c) *Area Feature (Polygons):* An area feature is a closed figure whose boundary encloses a homogenous area. These are represented by polygons defined by any natural boundaries. Examples of such features are water body, aquifer units, rock units, geomorphic units, etc.

Figure 2.2

Topological Data Elements in GIS

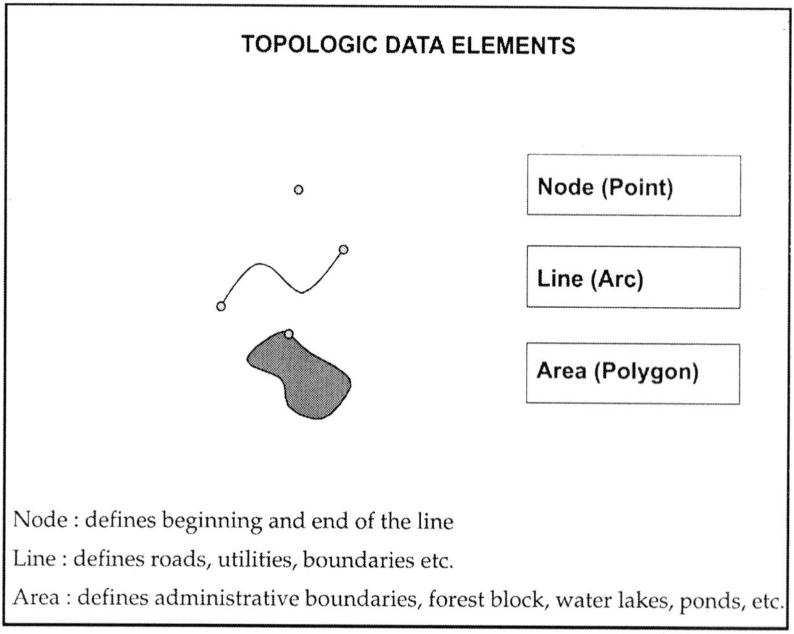

(ii) Database Models: There are two major types of models in GIS, as are represented in Figure 2.3, viz.: (a) Raster Model, and (b) Vector Model

Raster Data model: The raster model divides the entire study area into a regular grid of cells in specific sequence.

- The conventional sequence is row by row from top left corner.
- Each cell contains a single value.
- It is space filling since every location in the study area corresponds to a cell in the raster.

- One set of cells and associated values is a layer, and
- There may be many layers in a database, e.g., soil type, elevation, land use, land cover, etc.

Vector Data model: It reduces the entire study area into three geographical identities – the point, polygon and the area.

- Point, line and polygon can represent every geographical phenomenon, plus a label saying what it is. That the labels could be the actual names, or they could be numbers of the cross-reference.
- A map is a set of points; lines and areas that are defined both by their location in space with reference to a coordinate system and by their non-spatial attributes.
- Vector objects do not necessarily fill space; not all locations in space need to be referenced in the model.

Figure 2.3

Basic Data Models

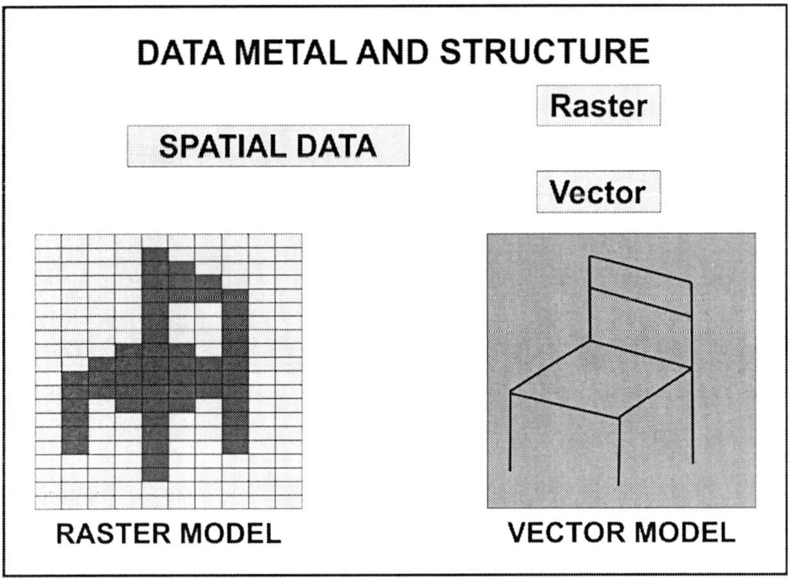

(iii) Database Management in GIS: A database here refers to a computerized collection of related information stored in such a way that retrieval can be performed by linking various pieces of information together. It consists of one or more data files, which are collection of related information treated as one unit on a computer. Databases are managed and accessed via a software termed Database Management System (DBMS). DBMS is a set of computer programs for organizing the information in a database. Typically, a DBMS contains routines for data input, verification, storage, retrieval and combination. The combination of hardware, software and the database itself is referred to as a database system.

The main characteristic of geographical database is its spatial nature. A spatial database is a collection of spatially referenced data that act as a model of reality. All the basic data types in geography/geology are spatially distributed such as geomorphological feature, rock type, well site, lineaments, roads, etc. Hence, Geographic Information System (GIS) provides an excellent tool to design, implement and manage geographical data in a most efficient manner.

Figure 2.4

Query and Analysis in GIS

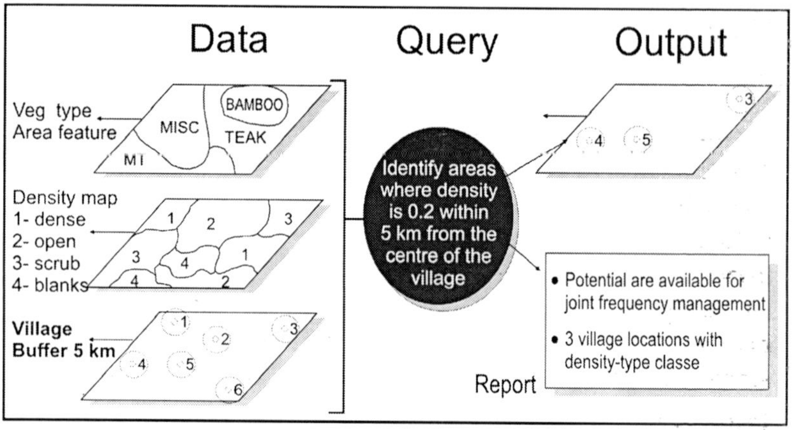

(iv) Query and Analysis in GIS: A corresponding database of thematic and statistics attributes is linked to the map features and can be sorted and classed, then queried to reveal important details not readily seen on the maps, as is shown in Figure 2.4. For example, population information relating to administrative units or municipalities might be sorted to show varying population densities among the municipalities. Such variations could then be shown graphically as charts or histograms on the map.

Advantages of GIS

The Geographic Information System (GIS) has been an effective tool for implementation and monitoring of municipal infrastructure. The use of GIS has been in vogue primarily due to the advantages mentioned below:

- Planning of project
- Make better decisions
- Visual analysis
- Improve organizational integration

Planning of Project

Advantage of GIS is often found in detailed planning of a project having a large spatial component, where analysis of the problem is a prerequisite at the start of the project. Thematic maps generation is possible on one or more than one base maps, example: the generation of a land use map on the basis of a soil composition, vegetation and topography. The unique combination of certain features facilitates the creation of such thematic maps. With the various modules within GIS it is possible to calculate surface, length, width and distance.

Making Decisions

The adage "better information leads to better decisions" is as true for GIS as it is for other information systems. A GIS, however, is not an automated decision-making system but a tool to query, analyse, and map the data in support of the decision-making process. GIS technology has been used to assist in tasks such as presenting information at planning, inquiries, helping resolve territorial disputes and siting pylons in such a way as to minimize visual intrusion.

Visual Analysis

Digital Terrain Modelling (DTM) is an important utility of GIS. Using DTM/3D modelling, landscape can be better visualized, leading to a better understanding of certain relations in the landscape. Many relevant calculations, such as (potential) lakes and water volumes, soil erosion volume (example: landslides), quantities of earth to be moved (channels, dams, roads, embankments, land levelling) and hydrological modelling becomes easier.

Not only in the previously mentioned fields but also in the social sciences, GIS can prove extremely useful. Besides the process of formulating scenarios for an Environmental Impact Assessment, GIS can be a valuable tool for sociologists to analyse administrative data such as population distribution, market localization and other related features. The driver for this closed-loop operation is the constant need for timely information about human activities and expectations concerning life in the real world. The specifics underlying those needs define the types and amounts of data information required. Once GIS users stipulate the specifics, they collect the data from multiform sources, such as already published maps and tabulations, current field observations, surveys, and aerial/satellite imagery. In the next step they convert the varied data into computer-compatible formats. The heart of the GIS operation lies within various techniques for analysis that users have devised as GIS evolved. They then present their reports, displays, new maps, statistics, and other kinds of computer-based, information-oriented products to decision makers. The test of value then happens by applying the results in the same real world that dictated the initial requirements. Data management through a GIS involves all of these facets.

Interfacing Remote Sensing and Geographic Information System

"Implementation of remote sensing and GIS has proven ideal for addressing the task of comprehending and combining all the components of ecosystems, human use of various resources, and changing human values in relation of natural resources. Remote Sensing and GIS can be used as tool to assist formulate the plan for the natural resource management in day-to-day activities".

GIS is fast becoming crucial to environmental monitoring, planning and impact assessment. It provides an excellent means of storing, retrieving, manipulating and analysing geographically referenced spatial information, usually stored in the form of a geographic database. Map information from standard sources is, however, updated only at set intervals, which makes the use of GIS somewhat limited, particularly in areas that are undergoing rapid dynamic change.

Remote sensing on the other hand, is an excellent tool as it allows repeated coverage of areas on a regular basis, making it possible to study changes over a period of time. Remote sensing data, e.g., are acquired and archived on a routine basis. Interfacing remote sensing-derived digital thematic maps with GIS thus provides a powerful mechanism not only to monitor environmental changes but also to permit the analysis of information derived from remote sensing in conjunction with a number of other thematic layers in the geographical database. It also allows for rapid updating of the geographic database, thus making the GIS a functional and flexible tool.

"The integration of remote sensing and GIS can be used to the full potential of these complementary technologies".

Thus, interfacing information derived from digital remote sensing with a GIS provides an excellent tool for monitoring environmental phenomenon.

References

Anuta, P.E. (1977), 'Computer Assisted Analysis Techniques for Remote Sensing Data Interpretation', *Geophysics*, Vol. 42.

Aronoff, S. (1989), *Geographic Information Systems: A Management Perspective*, WDL Publications, Ottawa, Canada.

Burrough, P.A. (1986), *Principles of Geographical Information Systems for Land Resources Assessment*, Clarendon Press, Oxford.

Curran, P.J. (1988), *Principles of Remote Sensing*, Longman, London.

ESRI (1993), 'ESRI Build National GIS for Real State Industries', *ARC News*.

Kennedy, M. and C.R. Meyers (1997), *Spatial Information System*, Introduction, Urban Studies Centre, Louisville, Kentucky.

Jensen, J.R. (1975), *Digital Image Processing*, Prentice-Hall, Englewood Cliffs, New Jersey.

Joseph, G. (2004), *Fundamentals of Remote Sensing*, Cambridge University Press, Cambridge, UK.

Reddy, M. Angi (2001), *Remote Sensing and Geographic Information System*, B.S. Publications, Hyderabad.

Sabins, J.R. (1997), *Remote Sensing Principles and Interpretation*, 3rd Edition, W.H. Freeman & Co., New York.

3

Digital Image Processing Techniques

Shankar Bagale and Ashwini Kumar Vaishnav

In general terms, an image is a digital picture or representation of an object. Remotely sensed image data are digital representations of the earth. The image consists of 2-dimensional array of pixels for each band. With the pixel it is possible to measure the density of the light-absorbing silver or dye deposited within a user-specified picture element in the image and converts this measurement into a digital brightness value, i.e., DN value (Digital number, pixel value). These representations form images when they are displayed on a screen or are output to a hardcopy. This pixel value represents the radiometric resolution of remote sensing data.

Each DN value is the output of the processes of analogue to digital conversion (A to D). The image display subsystems carry out the conversion operation, that of taking a digital quantity of an individual pixel value from an image. The surface of the ground is divided into parcels. Each parcel of land can be represented as a single pixel (picture element) on the image.

Image is a representation in the domain of digital elevation models (DEM) in which each pixel value can represent elevation.

Bands

Image data may include several bands of information. Each band is a set of data file values for a specific portion of the electromagnetic spectrum (as shown in Figure 3.1) of reflected light or some other user-defined

Figure 3.1

Pixels and Bands in a Raster Image

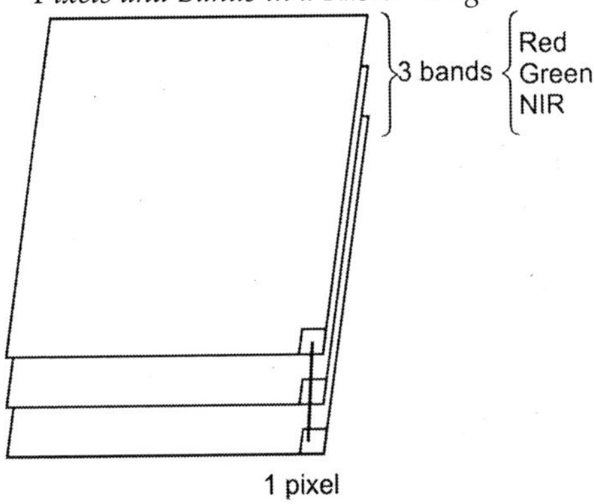

Figure 3.2

8-byte Gray Scale

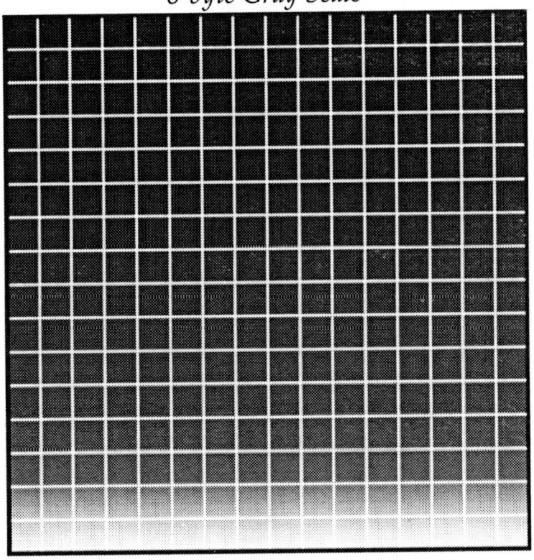

information created by combining or enhancing the original bands, or creating new bands from other sources. Every satellite will carry cameras with a group of bands such as red, green, blue, near-infrared (NIR), infrared, thermal, etc. Indian Remote Sensing Satellite (IRS) series carry Linear Imaging Self Scanner (LISS), Wide Field Sensor (WiFS), and the like cameras having green, blue, near-infrared, mid-infrared in each. This detail of band information of IRS is given later in this chapter. The satellite image and its band is shown in Figure 3.1.

Bands of image are occasionally referred to as layers. Once a band is imported into a GIS, it becomes a layer of information which can be processed in various ways. Additional layers can be created and added to the image file, such as layers created by combining existing layers.

This A to D conversion process results in a matrix of DN values that is usually recorded in 8-bits byte (values ranging from 0 to 255) for IRS. But in the other countries it will be recorded in 6, 10, 12-bits byte also (see the table), e.g., for gray band, i.e., in black and white its 8 bits colour appears as shown in Figure 3.2.

Table 3.1

Byte-wise Ranges of Colours

Quantization of Satellite (bits)	*Range*	*Values range*
6	$2^6 = 128$	0–127
8	$2^8 = 256$	0–255
10	$2^{10} = 512$	0–511
12	$2^{12} = 1028$	0–1027

The above gray band scale format is mostly used in most raw unprocessed satellite imagery and is stored and displayed in same format. This is also stored for a particular range of wavelength, i.e., for red, green, NIR, etc. A gray scale is a colour scale that ranges from black to white, with varying intermediate shades of gray. A commonly used gray scale for remote sensing image processing is a 256 shade gray scale, where a value of 0 represents a pure black colour, the value of 255 represents pure white, and each value in between represents a progressively darker shade of gray.

Display of Image

Computer will display the colour images using combination of RGB (Red, Green, and Blue). This RGB display is made up of a combination of red, green, and blue at varying levels of intensity, each ranging from 0–255. Each unique colour has its own combination of red, green, and blue levels. With all of the possible combinations of red, green, and blue values, this provides for a display system capable of using millions of different colours. In the diagram below, each unique colour is displayed with its red, green, and blue values.

The way the IRS data are mapped into three colours in the output image depends on the information that one wishes to be highlighted in the images The spectral characteristics of the target being observed and the type of information a researcher hopes to extract from the raw data determine which bands will be used in the composite and which colour (red, green, or blue) will be assigned to each band.

Figure 3.3

Coordinate System of RGB Display

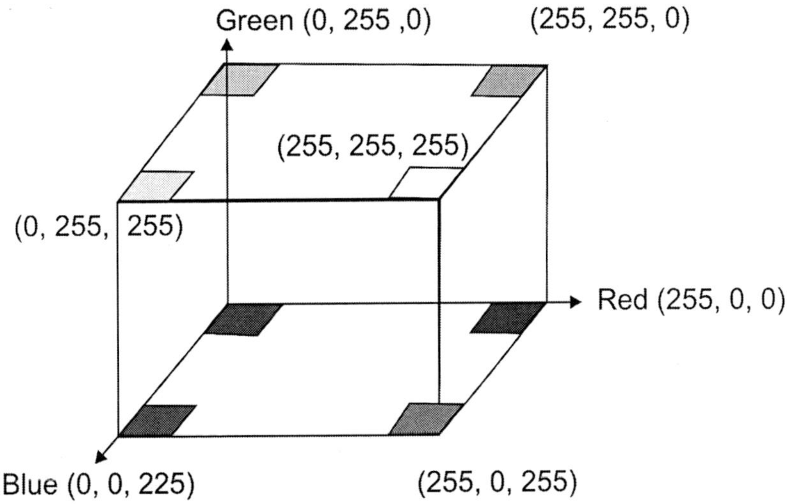

Regardless of the combination of bands used, the mapping of IRS sensor data to the RGB colour display is the same. Three bands are selected, each is assigned to one of the three primary RGB colours, and the value of each colour level is mapped to the measured value of each pixel in the appropriate band. For example, to create a composite image that maps the measured values of IRS bands 3, 2, and 1 to the colours red, green, and blue respectively, the colour of each pixel would be calculated using the following logic:

- The red value of the pixel would be set equal to the measured energy level of that pixel in band 3.
- The green value of the pixel would be set equal to the measured energy level of that pixel in band 2.

- The blue value of the pixel would be set equal to the measured energy level of that pixel in band 1.

This logic is repeated for every pixel in the scene being processed, until an entire image is produced with the pixel values derived from a combination of each of the three bands.

True-Colour Composite

True colour composite images are created by combining the Enhanced Thematic Mapper (ETM) spectral bands that most closely resemble the range of vision of the human eye. A true colour composite uses the visible red (band 3), visible green (band 2), and visible blue (band 1) channels to create an image that is very close to what a person would expect to see in a photograph of the same scene. The band to colour mapping for a 321 Composite are:

- Band 3 (Visible red) = red
- Band 2 (Visible green) = green
- Band 1 (Visible blue-green) = blue

True colour images are based entirely on reflected solar radiation in the visible portion of the electromagnetic spectrum. Haze in the atmosphere, shadows, clouds, and scattering – all affect the quality and usefulness of a true colour composite. True colour images are often low in contrast and hazy in appearance, since blue light is more easily scattered by the atmosphere.

True colour composite images can be very useful, especially when studying coastal regions, since energy in the visible bands can penetrate water surfaces. Particles in the water, such as sediment or algae, will reflect visible light and can therefore be detected by the visible sensors on the Landsat 7, ETM + sensor. Using true colour composite imagery, we can observe and measure the amount of sediment flowing from rivers into larger bodies of water such as the Chesapeake Bay following storm events. We can also locate and measure large blooms of algae that threaten the water quality and fishery production in coastal waterways.

False Colour Composite (Near Infrared Composite)

The human eye can only distinguish between about 16 shades of gray in an image, however it is able to distinguish between millions of different colours. Thus, a common image enhancement technique is to assign specific digital number (DN) values (or ranges of DN values) to specific colours, thereby increasing the contrast of particular DN values with the surrounding pixels in an image. An entire image can be converted from a gray scale to a colour image, or portions of an image that represent the DN values of interest can be coloured.

A Near Infrared composite eliminates the visible blue band and uses a Near Infrared (NIR) band to produce the image. The resulting composite does not resemble what the human eye will see (for example, vegetation is red instead of green); however it is very useful to researchers. The mapping of colour to band is:

Band 4 (NIR) = red

Band 3 (Visible red) = green

Band 2 (Visible green) = blue

Thus, in a 432 NIR composite image, vegetation is vividly depicted as varying shades of red. Since different types of vegetation have different levels of chlorophyll in their leaves, each type of plant has its own shade of red. This makes a 432 composite very useful in determining the extent of vegetation and in classifying different vegetation types as seen from space.

Water, which absorbs nearly all of the NIR energy that reaches its surface, appears very dark, nearly black, in a 432 NIR composite image. Therefore, this type of imagery would not be useful for studying underwater features.

Colour Composite

A true colour image is one for which the colours have been assigned to DN values that represent the actual spectral range of the colours used in the image (blue features appear blue, green features appear green, red features appear red, etc). A photograph is an example of a true colour image. Generally the red pixels in images are assigned to pixels that have the strongest spectral response in the red band of the visible light spectrum, and the same is true for the green and the blue colours. The result is an image that most closely represents the earth's surface as it would appear to the human eye.

False colour is a technique by which colours are assigned to spectral bands that do not equate to the spectral range of the selected colour. This allows an analyst to highlight particular features of interest using a colour scheme that makes the features stand out.

Functions of Image Processing

The various functions of image processing are as follows:

- Preprocessing – Radiometric correction, Geometric correction, etc.
- Display and enhancement:
  - Image display in black/white (B/W), True or False colour composite (FCC).
  - Image algebra.
  - Edge enhancement.
  - Magnification, reduction, roam, pan.
  - Digital Elevation Model (DEM).
  - Animation
  - Image compression, etc.
- Information Extraction:
  - Supervised classification.
  - Unsupervised classification.
  - Radar image processing.
  - Merging and sub-setting, etc.
- Image and map composition.
- Integrating image processing with GIS.

Preprocessing

Preprocessing is an important and diverse set of image preparation programs that act to offset problems with the band data and recalculate DN values that minimize these problems. Among the programs that optimize these values are atmospheric correction (affecting the DNs of surface materials because of radiance from the atmosphere itself, involving attenuation and scattering); sun illumination geometry; surface-induced geometric distortions; spacecraft velocity and attitude variations (roll, pitch, and yaw); effects of earth rotation, elevation, curvature (including skew effects), abnormalities of instrument performance (irregularities of detector response and scan mode such as variations in mirror oscillations); loss of specific scan lines (requires de-striping), and others. Once performed on the raw data, these adjustments require appropriate radiometric and geometric corrections.

Before digital images can be analysed, they usually require some degree of preprocessing. This may involve radiometric corrections, which attempt to remove the effects of sensor errors and/or environmental factors. Common corrections of this type include those that attempt to adjust DN values that have been affected by atmospheric interference or absorption. Ancillary data collected at the same time the image was obtained can be used as a calibration tool to support radiometric corrections.

Sensor Normalization

Problems of cloudiness and unsuitability of acquisition date, and analysis of historic data may necessitate the use of data from different sensors. Response characteristics of sensors differ to the extent that simple radiance calibration is not sufficient for getting actual spectral indices while using data from more than one sensor. By using common sites located on different sensor data sets of the same date, spectral indices may be obtained.

The satellite sensor provides spectral data in terms of band-wise digital numbers. In order to normalize the satellite data, it is necessary to convert these numbers into spectral radiance values. This conversion requires calibration constants, viz., band-wise offset and gain values of different sensors. Based on the performance of sensor aboard the satellite, these offset and gain values are changed accordingly. Hence the operational offset and gain values of the sensors supplied by National Data Centre (NDC), National Remote Sensing Agency (NRSA) are used in normalizing the satellite data of various sensors. Thus, normalized (radiance images) are further used in the analysis. Data was generally expressed in radiance values. Gain and bias values are obtained from NDC for various data sets. However, in some operations it was more feasible to express data in digital numbers. Conversion of digital numbers to radiance is done using the following formulae

$$L(\lambda) = \frac{L(\lambda)_{max} = L(\lambda)}{DN_{max}} DN(\lambda) + L(\lambda)_{min}$$

$L(\lambda)$ = spectral radiance (m W cm^{-2} sr^{-1} μm^{-1})

$L(\lambda)_{max}$ = maximum spectral radiance (m W cm^{-2} sr^{-1} μm^{-1}) registered by the sensor

$L(\lambda)_{min}$ = minimum spectral radiance (m W cm^{-2} sr^{-1} μm^{-1}) registered by the sensor

DN = digital number, absolute calibrated by the distributor

Dn_{max} = maximum digital number

Geometric Correction

Geometric corrections are also a very important form of preprocessing. This method attempts to rectify any error introduced into an image by the geometry of the curved earth's surface and its rotation and the movement of the satellite. There are several other reasons for rectifying image data:

- comparing pixels scene to scene in applications
- developing GIS data bases for GIS modelling
- identifying training samples according to map coordinates prior to classification
- creating accurate scaled photomaps
- overlaying an image with vector data, such as Arc Info
- comparing images that are originally at different scales
- extracting accurate distance and area measurements
- mosaicking images

Before rectifying the data, you must determine the appropriate coordinate system for the data base. To select the optimum map projection and coordinate system, the primary use for the data base must be considered. If you are doing a government project, the projection may be predetermined. A commonly used projection in the United States government is State Plane. Use an equal area projection for thematic or distribution maps and conformal or equal area projections for presentation maps. Before selecting a map projection, consider the following:

- How large or small an area is mapped? Different projections are intended for different size areas.
- Where on the globe is the study area? Polar regions and equatorial regions require different projections for maximum accuracy.
- What is the extent of the study area? Circular, north-south, east-west, and oblique areas may all require different projection systems (ESRI, 1992).

A map projection system is any system designed to represent the surface of a sphere or spheroid (such as the earth) on a plane. There are a number of different map projection methods. Since flattening a sphere to a plane causes distortions to the surface, each map projection system compromises accuracy between certain properties, such as conservation of distance, angle, or area. However, to maintain equal area, the shapes, angles, and scale in parts of the map may be distorted (Jensen, 2004). There are a number of map coordinate systems for determining location on an image. These coordinate systems conform to a grid, latitude/longitude Lat/Lon and are expressed as X, Y (column, row) pairs of numbers. Each map projection system is associated with a map coordinate system. Lat/Lon is a spherical coordinate system that is not associated with a map projection. Lat/Lon expresses locations in the terms of a spheroid, not a plane. Therefore, an image is not usually rectified to Lat/Lon, although it is possible to convert images to Lat/Lon.

Geometric correction is a process by which points in an image are registered to corresponding points on a map or other image that has already been rectified. The goal of geometric correction is to put image elements in their proper planimetric (x and y) positions. Hence there is a need to geometrically correct the image so that it can be represented on a planar surface and conforms to other images and has the integrity of a map. Rectification is performed by transforming the raster data from one grid system to the other using first or higher order polynomial. Since the pixels of the new grid may not align with the pixels of the original grid, the pixels must be resampled.

Resampling is the process of extrapolating data values for the pixels on the new grid from the values of the source pixels. Resampling is one approach commonly used to produce better estimates of the DN values for individual pixels; the net effect is that the resulting redistribution of pixels involves their spatial displacements to new, more accurate relative positions. However, the radiometric values of the displaced pixels no longer represent the real world values that would be obtained if this new pixel array could be resensed by the scanner (this situation is alleviated somewhat if the sensor is a Charge-Coupled Device. The particular mixture of surface objects or materials in the original pixel has changed somewhat (depending on pixel size, number of classes and their proportions falling within the pixel, extent of continuation of these features in neighbouring pixels [a pond may fall within one or just a few pixels; a forest can spread over many contiguous pixels]). In simple words, the corrections have led to a pixel that at the time of sampling covered ground A being shifted to a position that have A values but should if properly located represent ground B.

An estimate of the new brightness value (as a DN) that is closer to the B condition is made by some mathematical resampling technique. Three sampling algorithms are commonly used:

In the Nearest Neighbour technique, the transformed pixel takes the value of the closest pixel in the pre-shifted array. Nearest neighbour is the only appropriate resampling method for thematic files, which may be a drawback in some applications.

In the Bilear Interpolation approach, the average of the DNs for the 4 pixels surrounding the transformed output pixel is used.

The Cubic Convolution technique averages the 16 closest input pixels; this usually leads to the sharpest image.

Registration

In many cases, images of one area that are collected from different sources must be used together. To be able to compare separate images pixel by pixel, the pixel grids of each image must conform to the other images in the data base. The tools for rectifying image data are used to transform disparate images to the same coordinate system. Registration is the process of making an image conform to another image. A map coordinate system is not necessarily involved.

Georeferencing

Georeferencing refers to the process of assigning map coordinates to image data. The image data may already be projected onto the desired plane, but not yet referenced to the proper coordinate system. Rectification, by definition, involves georeferencing, since all map projection systems are associated with map coordinates. Image-to-image registration involves georeferencing only if the reference image is already georeferenced. Georeferencing, by itself, involves changing only the map coordinate information in the image file. The grid of the image does not change. Geocoded data are images that have been rectified to a particular map projection and pixel size, and usually have had radiometric corrections applied. It is possible to purchase image data that is already geocoded. Geocoded data should be rectified only if they must conform to a different projection system or be registered to other rectified data.

Rectification Steps

Rectification is the conversion of data file coordinates to some other grid and coordinate system, called a reference system. Rectifying or registering image data on disk involves the following general steps, regardless of the application:

(i) Locate Ground Control Points (GCPs).

(ii) Compute and test a transformation.

(iii) Create an output image file with the new coordinate information in the header.

Ground Control Points

GCPs are specific pixels in an image for which the output map coordinates (or other output coordinates) are known. GCPs consist of two X,Y pairs of coordinates:

- source coordinates – usually data file coordinates in the image being rectified
- reference coordinates – the coordinates of the map or reference image to which the source image is being registered.

The term map coordinates is sometimes used loosely to apply to reference coordinates and rectified coordinates. These coordinates are not limited to map coordinates. For example, in image-to-image registration, map coordinates are not necessary.

GCPs for large-scale imagery might include the intersection of two roads, airport runways, utility corridors, towers, or buildings. For small-scale imagery, larger features such as urban areas or geologic features may be used. Landmarks that can vary (e.g., the edges of lakes or other water bodies, vegetation, etc.) should not be used.

Polynomial Transformation

Polynomial equations are used to convert source file coordinates to rectified map coordinates. Depending upon the distortion in the imagery, the number of GCPs used, and their locations relative to one another, complex polynomial equations may be required to express the needed transformation. The degree of complexity of the polynomial is expressed as the order of the polynomial. The order is simply the highest exponent used in the polynomial. The order of transformation is the order of the polynomial used in the transformation. A correlation threshold is used to accept or discard points. The correlation ranges from -1.000 to +1.000. The threshold is an absolute value threshold ranging from 0.000 to 1.000. A value of 0.000 indicates a bad match and a value of 1.000 indicates an exact match. Values above 0.8000 or 0.9000 are recommended. If a match cannot be made because the absolute value of the correlation is less than the threshold, you have the option to discard points.

Transformation Matrix

A transformation matrix is computed from the GCPs. The matrix consists of coefficients that are used in polynomial equations to convert the coordinates. The size of the matrix depends upon the order of transformation. The goal in calculating the coefficients of the transformation matrix is to derive the polynomial equations for which there is the least possible amount of error when they are used to transform the reference coordinates of the GCPs into the source coordinates.

Linear Transformations

A first order transformation is a linear transformation. It can change:

- location in X and/or Y
- scale in X and/or Y
- skew in X and/or Y
- rotation

First-order transformations can be used to project raw imagery to a planar map projection, convert a planar map projection to another planar map projection, and when rectifying relatively small image areas. You can perform simple linear transformations to an image displayed in a viewer or to the transformation matrix itself. Linear transformations may be required before collecting GCPs on the displayed image.

Non-linear Transformations

Transformations of the second order or higher are non-linear transformations. These transformations can correct non-linear distortions. The process of correcting non-linear distortions is also known as rubber sheeting. Second-order transformations can be used to convert Lat/Lon data to a planar projection, for data covering a large area (to account for the earth's curvature), and with distorted data (for example, due to camera lens distortion).

Third-order transformations are used with distorted aerial photographs, on scans of warped maps and with radar imagery. Fourth-order transformations can be used on very distorted aerial photographs. The transformation matrix for a transformation of order t contains this number of coefficients: It is multiplied by two for the two sets of coefficients – one set for X, one for Y. An easier way to arrive at the same number is: Clearly, the size of the transformation matrix increases with the order of the transformation.

Classification

Classification is a part of image processing for identifying groups of pixels that have similar spectral characteristics and to determine the various features or land cover classes represented by these groups. Therefore, it is also called as a process of information extraction. Visual classification relies on the analyst's ability to use visual elements (tone, contrast, shape, etc.) to classify an image. Digital image classification is based on the spectral information used to create the image and classifies each individual pixel based on its spectral characteristics. The result of a classification is that all pixels in an image are assigned to particular classes or themes (e.g., water, coniferous forest, deciduous forest, corn, wheat, etc.), resulting in a classified image that is essentially a thematic map of the original image. The theme of thc classification is selectable, thus a classification can be performed to observe land use patterns, crop patterns, vegetation types.

While classifying an image one must distinguish between spectral classes and information classes. Spectral classes are groups of pixels that have nearly uniform spectral characteristics. Information classes are the various themes or groups the analyst is attempting to identify in an image. Information classes may include such classes as deciduous and coniferous forests, various agricultural crop types, or inland bodies of water. The objective of image classification is to match the spectral classes in the data to the information classes of interest.

Though any image can be classified, multi-spectral imagery tends to be used most often. One band classification is usually very difficult to classify since more than one surface type will exhibit the same digital number. Thus, any spectral classe in a single band classification will likely contain several information classes, and distinguishing between them would be difficult. Normally two or more bands are used for classification, and their combined digital numbers are used to identify the spectral signatures of the spectral classes present in the image. The more bands used to create a classification, the more likely that the analyst will get a set of unique land cover classes.

Classification is another tool that is very useful when multi-spectral imageries of the same geographical region is compared. Algorithms can be used that derive a value for each pixel in the image from its brightness values in each image. Plotting the resulting data on a 2 or 3 dimensional graph can identify clusters of pixels that share common spectral characteristics across multiple bands. There are two approaches for classification:

(i) Unsupervised classification.
(ii) Supervised classification.

Unsupervised Classification

In unsupervised classification, the objective is to group multi-band spectral response patterns into clusters that are statistically separable. Thus, a small range of digital numbers for say 3 bands can establish one cluster that is set apart from a specified range combination for another cluster (and so forth). Separation will depend on the parameters chosen to differentiate. Visualize this process with the Figure 3.4 for four classes: A = Agriculture; D= Desert; M = Mountains; W = Water. For separation of these classes on a computer, each pixel in an image is assigned to one of the clusters as being most similar to it in DN combination value. Generally, in an area within an image, multiple pixels in the same cluster correspond to some (initially unknown) ground feature or class so that patterns of gray levels result in a new image depicting the spatial distribution of the clusters. These levels can then be assigned colours to produce a cluster map. The trick then becomes one of trying to relate the different clusters to meaningful ground categories. Doing this by either being adequately familiar with the major classes expected in the scene, or, where feasible, by visiting the scene (ground truth) and visually correlating map patterns to their ground counterparts; since the classes are not selected beforehand.

Figure 3.4

Clustering of Particular Class from Multi-spectral Satellite Image

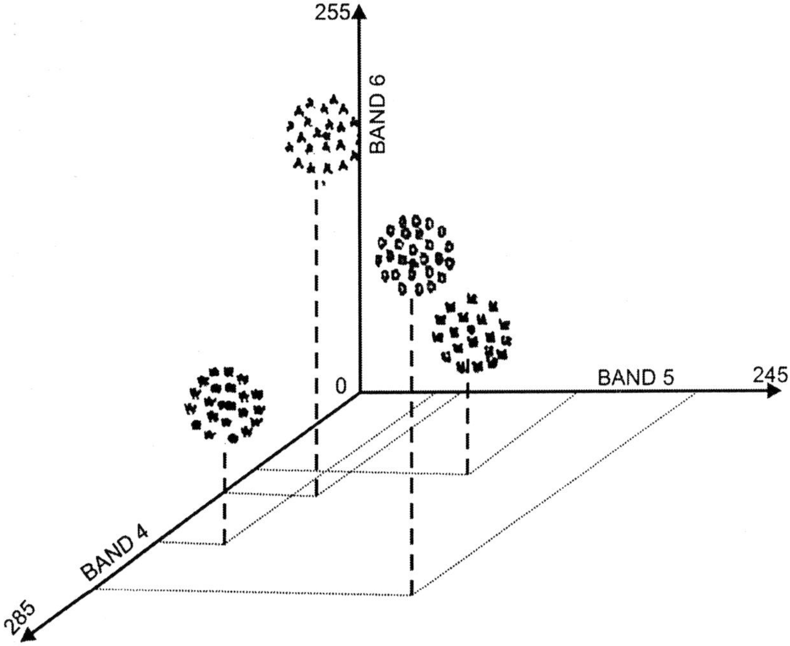

Algorithms calculate the cluster values from these bands such as k-mean, Iso data. It automatically determines the maximum number of clusters by the parameters selected in the processing. This process typically has the effect of producing so many clusters that the resulting classified image becomes too cluttered and, thus, more difficult to interpret in terms of assigned classes.

Unsupervised classification is too much of a generalization and the clusters only roughly match some of the actual classes. Its value is mainly as a guide to the spectral content of a scene to aid in making a preliminary interpretation prior to conducting the much more powerful supervised classification procedures.

Supervised Classification

The principles behind supervised classification are considered in more detail. The fact that the pixel DNs for a specified number of bands are selected from areas in the scene that are a priori of known identity, i.e., can be named as classes of real features, materials, etc., allows establishment of training sites that become the basis of setting up the statistical parameters used to classify pixels outside these sites. The process by which training sites were selected for the Morro Bay scene, together with a plot of spectral signatures and a table of statistical means and site sample sizes illustrate the preparatory phase involved in this mode of classification.

Supervised classification is much more accurate for mapping classes, but depends heavily on the cognition and skills of the image specialist. The specialist must recognize conventional classes (real and familiar) or meaningful (but somewhat artificial) classes in a scene from prior knowledge of the geographical region and the surface cover types present in the image or by on-site visits. This familiarity allows the specialist to choose and set up discrete classes (thus supervising the selection) and then, assign them category names. The specialists also locate training sites on the image to identify the classes.

Training sites are areas representing each known land cover category that appear fairly homogeneous on the image (as determined by similarity in tone or colour within shapes delineating the category). Specialists locate and circumscribe them with polygonal boundaries drawn (using the computer mouse) on the image display. For each class thus outlined, mean values and variances of the DNs for each band used to classify them are calculated from all the pixels enclosed in the site. More than one polygon can be established for any class. When DNs are plotted as a function of the band sequence (increasing with wavelength), the result is a spectral signature or spectral response curve for that class. In reality the spectral signature is for all of the materials within the site that interact with the incoming radiation.

Statistical stage: Once the training sites have been established, the numerical information in the entire image's spectral bands are used to define the spectral "signature" of each class. Once the computer has determined the signatures for each class, it will compare every pixel to the signatures and label it as the class that it is mathematically closest to. Taking into consideration the identified training site mean standard deviation, covariance and correlation matrix is calculated for each training sites. Classification now proceeds by statistical processing in which every pixel is compared with the various signatures and assigned to the class whose signature comes closest. (A few pixels in a scene do not match and remain unclassified, because these may belong to a class not recognized or defined).

Classification stage: Thus, in a supervised classification, the analyst starts with information classes and uses these to define spectral classes. Each pixel in the image is then assigned to the class which it most closely resembles. This stage includes various classifiers like:

(i) Minimum distance to mean classifier: It calculates the value of mean for each training site in each band, which forms the mean vector for each training class. The classifier then calculates the distance to each mean vector. In each pixel it vectors and assigns the pixel to the class to which distance is minimum. This classifier is not suitable for application where spectral classes are close to one another and have high variance, because only mean is considerable and it is insensitive to variance.

(ii) Parallel piped classifier: It considers the range of values in each training site, i.e., the minimum and maximum of each class is considered. A pixel is classified as per the category range.

(iii) Maximum likelihood classifier: It is the most widely used classifier and takes into account both variance and covariance. The assumption is made that the distribution of points follow a Gaussian pattern, i.e., the response can be categorized by mean vector and covariance matrix. The probability of each pixel belonging to each class is calculated and the pixel is assigned to the class for which probability will be maximized.

Normalized Difference Vegetation Index (NDVI)

It is the ratio of difference in NIR & Red and addition of NIR & Red. NDVI is commonly used to describe vegetation characteristics. NDVI provides an estimate of vegetation health and a means of monitoring changes in vegetation over time. NDVI from IRS, LISS-III images are calculated as the ratio:

$$NDVI = \frac{(LISS\,band\,4) - (LISS\,band\,3)}{(LISS\,band\,4) + (LISS\,band\,3)}$$

Reasons for considering NIR and Red bands are explained in Chapter 2. The NDVI value is used for classification of paddy. IRS, LISS band 4 represents reflection in the Near Infrared (NIR) portion (directly related to green biomass), and LISS band 3 represents the reflection in the Red portion. NDVI values change during the growth cycle as they are directly related to plant height and biomass.

Land Use and Land Cover (LULC) Classification

Land use refers to man's activities and the various uses which are carried on the land like built-up lands, urban, cultivated agriculture, etc., while land cover refers to natural coverage on the earth's surface like natural vegetation, forests, water bodies, etc. Although land use is generally inferred based on the cover, yet both the terms are related and interchangeable.

LULC classification means information extracted of land use/land cover of the earth surface in various categories such as agriculture, urban, etc. This classified information plays vital a role in planning and management of various developmental activities like agriculture, water resource projects, irrigation management, command area development and also for conservative measures to checking and analysing damage likely to be caused by various natural disaster. LULC classification is very essential for resource mapping, management and conservation activities. Land use data also helps in the analysis of environmental conditions and the associated problems.

For LULC classification satellite images are used because satellite images provide a synoptic view of a large part of earth surface in near real time depicting accurate picture of land use and land cover. Land use pattern generally reflects the extent of resource utilization and indicates the productivity of the area. Therefore, inventory of land use/land cover categories, their spatial distribution and understanding of the changes in their pattern.

The LULC classification of the study area is used for this study of planning for command area and submergence area assessment. It is also used for canal fixing alignment.

Description of Land Use/Land Cover Classes

(i) *Built-up land*: It is an area of human habitation developed due to non-agricultural use. All man-made constructions, covering the land surface come in this category. These are human settlements comprising residential areas, transportation and communication lines, industrial and commercial complexes, utilities and services, etc. Collectively, cities and towns (Urban) and villages (Rural) are included under this category.

(ii) *Agricultural land*: It is defined as the land primarily used for farming and for production of food grains, fibre and other commercial and horticultural crops. It includes cropland (irrigated and un-irrigated), fallow, plantations, aquaculture, etc.

(a) *Crop land*: These are the area with standing crop as on date of satellite overpass. They are widely distributed in India, and prominently appear in the irrigated areas irrespective of the source of irrigation. Three cropping seasons are found in India: *Kharif* (June/July – September/October), *Rabi* (November/December-February/March), and *Zaid* (April/May).

(b) *Fallow lands*: These are the lands which are taken up for cultivation but are temporarily allowed to rest, uncropped for one or more seasons, but less than one year.

(c) *Plantation*: It is an area under agricultural tree crops planted, adopting certain agricultural management techniques.

(iii) *Forest*: These are the areas bearing an association predominantly of trees and other vegetation types (within the notified forest boundaries as in toposheets) capable of producing timber and other forest products.

(a) *Evergreen/semi-evergreen forest*: A forest, which comprises of thick canopy of tall trees and predominantly remains green throughout the year, is called evergreen forest. It includes both coniferous and tropical broad leaf evergreen trees. Semi-evergreen is a mixture of both deciduous and evergreen trees, but the latter is predominating.

(b) *Dense forest*: A forest area where the vegetative crown density is more than 40 per cent of the canopy cover comes under dense forest.

(c) *Open forest*: A forest area where the vegetative crown density is between 10 per cent and 40 per cent of the canopy cover comes under open forest.

(d) *Degraded/scrub forest:* This is a forest area where scrub is dominant. Scrub is a stunted tree or bush.

(e) *Forest plantations*: It is an area of trees of species of forestry importance and raised in forest area. These forest areas are artificially planted with trees, either in the open spaces or by clearing already existing forests of economically inferior species.

Table 3.2

Land Use/Land Cover Classification

S. No.	*LULC category*	*LULC 1st Level*	*LULC 2nd Level*
1	Built-up land	Urban (Town)	Residential Commercial Recreational Educational Public & Semi-public, etc.
2	Agricultural land	Crop Land	*Kharif* crop *Rabi* crop *Zaid* crop
		Fallow Lands	Current and Long
		Plantation	Agriculture Horticulture Aquaculture
3	Grass lands		
4	Forests	Evergreen	Dense/Closed Open
		Deciduous	Closed Open
		Forest Plantation	
		Scrub Forest	
		Blank, etc.	
5	Wastelands	Salt Affected	Slight Moderate Strong
		Gullied/Ravinous	Gullied Shallow Deep
		Scrub	Dense Open
		Sandy	Desert Coastal Riverside
		Stony/Rocky	
6	Wetlands	Natural Man-made, etc.	
7	Water Bodies	River/Streams Lake/Pond/Tank/Reservoir	
8	Snow		
9	Other		

(iv) *Wastelands*: These are degraded lands which can be brought under vegetative cover, with reasonable effort and which are currently underutilized, and lands which are deteriorating due to lack of appropriate water and soil management or on account of natural causes. Barren rocky/ Stony waste/Sheet rock areas category are only observed in the study area.

  (a) *Land with scrub*: This land is generally prone to deterioration due to erosion and being covered with scrubs.
  (b) *Land without scrub*: This land is also prone to deterioration due to erosion and lack of vegetative cover.
  (c) *Barren rocky/Stony waste/Sheet-rock area*: These areas are defined as rock exposures of varying lithology, often barren and devoid of soil cover and vegetation.

(v) *Water bodies*: This class comprises areas of surface water, either impounded in the form of tanks, ponds, lakes and reservoirs or flowing as streams, rivers, canals, etc. Cooling pond, cooling reservoir, abandoned quarries with water bay also comes under this category, but they are not present in the study area.

  (a) *River*: It is a natural course of water flowing upon the land surface along a definite channel.
  (b) *River sand*: It is the stabilized accumulation of sand in the river.
  (c) *Lakes/Pond*: A lake is a large body of surface impounded water, natural or artificial within the landmass. Reservoir: A reservoir is an artificial lake created by construction of a dam across the river specifically for the generation of hydroelectric power or irrigation or water supply for domestic/industrial uses or flood control or in combination of these.
  (d) *Tank*: Tanks are small lakes or impounded water.

Image Interpretation Techniques

Definition

Image interpretation is defined as 'the art of examining images for the purpose of identifying objects or surface features and judging their significance. Interpreters study remotely sensed data and attempt through logical processes in detecting identifying, classifying, measuring and evaluating the significance of physical and cultural objects, their patterns and spatial relationships. Image interpretation of remotely sensed data can be attempted either by visual or digital techniques of analysis. The purpose of applying either of the above two techniques is for feature identification and classification.

Aspect of Image Interpretation

Image interpretation is a complex process of physical, physiological activities occurring in a sequence of time. The sequence begins with the detection and identification of images and later by their measurements. Images are then considered in terms of information, usually non-pictorial and finally deductions are made. The various aspects of image interpretation are listed below in a simpler form. Most of these have overlapping functions.

(i) *Detection:* It is a process of 'picking out' an object or element from photo or image through interpretation techniques. It may be detection of point or line locations, e.g., agricultural field or small settlement.

(ii) *Recognition and identification:* It is a process of classification or trying to distinguish an object by its characteristics or patterns which are familiar on the image. It precedes the process of detection. Sometimes it is also termed as photo-reading, e.g., water features: stream, canal, tank, sand, etc.

(iii) *Analysis:* It is a process of resolving or separating a set of objects or features having similar set of characters. In analysis, 'lines of separation' are drawn between group of objects and the degree of reliability of these lines can also be indicated, e.g., sands as that of river, desert, coastal, etc.

(iv) *Classification:* It is a process of identification and grouping of objects or features resolved by analysis. It arranges 'features of recurrence' in the same class or group to which the feature belongs. Any wrong identification and analysis often may lead to missclassification.

(v) *Deduction:* It is a process wherein, inferences are drawn about the objects based on direct or indirect evidence of the information or phenomena under study, e.g., salt encrustation inference; agricultural activity. Deductions may be firmly confirmed by ground checks to avoid missclassification.

(vi) *Idealization:* It is a process of drawing ideal or standard representation from what is actually identified and interpreted from the image or map, e.g., set of symbols or colours to be adopted in wasteland maps or an ideal example on ravines. This process helps in developing image interpretation key.

Element of Image Interpretation

There are certain fundamental photo-elements or image characteristics seen on an image which aid in visual interpretation of satellite imagery. Although, there is, a difference of opinion on the number of elements to be included, however there is a general consensus on the following:

(i) *Tone or Colour:* Different surface objects reflect and emit different amounts of radiant energy. These differences are recorded as tonal/colour or density variations on the imagery. In black and white images, objects are appearing in different gray tones. These gray tones often fail to provide the interpreter a clear perception of objects. Whereas, true colour or false colour imagery increase interpretability by providing a subtle tonal contrast between them. Tonal contrast can be enhanced or reduced optically or by enhancement techniques using computers.

(ii) *Pattern:* It refers to the spatial arrangements of surface features which are characteristic of both natural and man-made objects. Similar features under similar environmental conditions reflect similar patterns of recurrence, e.g., salt affected land in irrigated areas and gullies in eroded areas. Pattern is of several types, e.g., linear – road/rail canal; linear-streams, creeks; contiguous-snow, sand; clustered-settlements; dispersed-forest blanks, salt affected patches; regular – orchards, strip cropping, etc. More often, patterns also reflect associations, e.g., intensity of drainage patterns shows its relation with rock types, soil texture, rainfall run-off, etc.

(iii) *Texture:* It is defined as a repetition of a basic pattern. Texture in the image is due to tonal repetitions in a group of objects which are often too small to be discernible. It creates a visual impression of surface roughness or smoothness of objects and is a useful photo element in image interpretation, e.g., salt affected land – fine to mottle; rocky areas – coarse.

(iv) *Size:* It refers to the spatial dimension of object on ground. Size of an object is a function of scale of the image or photo and is also measurable. There are different objects with varying sizes and shapes, e.g., gullies – varying size and depth; sandy areas – large and extensive; hump areas – small in extent.

(v) *Shape:* It refers to physical form of an object and is also a function of scale of the image or photo. Size and shape are interrelated. In the image, shape refers to plan or top view of the object, as seen by the satellite. Shape can be irregular, e.g., salt affected patches; boundary of undulating uplands; regular and uniform – snow or glacier.

(vi) *Shadow:* They are cast due to sun's illumination angle; size and shape of the object are sensor viewing angle. The shape and profile of shadows help in identifying different surface objects, e.g., clouds, nature of hill slopes' aspect, apparent relief, etc. They also help in arriving at tree heights or chimney/building heights on aerial photos.

(vii) *Location:* The geographical site and location of the object often provide clue for identifying objects and understanding their genesis, e.g., salt affected land, inland river/desert plains, hump or forest blanks – hill slopes; snow or glacier – mountain peaks, etc.

(viii) *Association:* It refers to situation of the object with respect to other surface and neighbouring features, e.g., canals with agricultural fields; marsh or swamps, flood plains and tidal flats; gullies or ravines with severely eroded lands.

(ix) *Resolution:* It is of two types, spatial and spectral. The former refers to 'picture element' or pixel discernible on the image or smallest area resolvable or identifiable on ground. Spatial resolution allows the interpreter to detect and distinguish the smallest object on the ground. On IRS P6 LISS IV image the smallest object which can be resolvable is 5.8 m on the ground.

References

Burrough, P.A. (1996), *Principles of Geographic Information Systems for Land Resources Management*, Clarendon Press, Oxford.

Chrisman, N.R. (1997), *Exploring Geographic Information Systems*, Wiley, New York.

Curran, P. (1989), *Principles of Remote Sensing*, Longman, London.

Date, C.J. (1995), *An Introduction to Database Systems*, VI Edition, Addison-Wesley, Reading, Mass.

Gonzalez, R.C. and P. Wintz (1983), *Digital Image Processing Addition*, Chapter 4, Wesley Publishing Company, California, pp. 140–205.

Jensen, J.R. (1986), *Introduction to Digital Image Processing*, Chapter 7, Prentice-Hall, Englewood Cliffs, New Jersey, pp. 117–76.

—— (2004), *Remote Sensing of the Environment: An Earth Resource Perspective*, University of South Carolina, Columbia, USA.

Joseph, G. (2004), *Fundamentals of Remote Sensing*, Cambridge University Press, Cambridge, UK.

Lawrini, R. and D. Thompson (1992), *Fundamentals of Spatial Information Systems*, Academic Press, London.

Lillesand, T.M. and R.W. Keifer (1979, 1994 and 2000), *Remote Sensing and Image Interpretation*, John Wiley & Sons, New York.

NRSA (1995, 1999), *IRS Handbook*, National Remote Sensing Agency, Hyderabad.

Reddy, M. Angi (2001), *Remote Sensing and Geographic Information System*, B.S. Publications, Hyderabad.

Tomlin, C.D. (1990), *Geographic Information Systems and Cartographic Modeling*, Prentice-Hall, Englewood Cliffs, New Jersey.

Vyas, P.R. (1996), *Principles of Geographic Information Systems for Land Resources Management*, Clarendon Press, Oxford.

Walsh, S.J. (1987), 'Comparison of NOAA-AVHRR Data to Meteorological Drought Indices', *Photogrammetric Engineering and Remote Sensing*, Vol. 53, pp. 1069–74.

Part – 2

Application of Remote Sensing and GIS for Natural Resources Management

4

Land Use/Land Cover Change Detection Studies Using Remote Sensing and GIS Techniques

The Case of Dhari Taluka, Gujarat

Jagdish Soni and R.P. Naraniwal

Land is the sole source of the sustenance of mankind, supporting the plants and animal on it providing the food, fibre and shelter. The growing pressure of population coupled with and increasing variety of demands created extra pressure on the available resources all over the country. Hence it is a necessary prerequisite to know the present or existing use of land in order to plan for its optimal use.

The natural resources (vegetation, water and soil) are responsive to human interaction and these together with terrain features determine the selection of proper land use pattern, which also in someway reflects the cultural, social and economic conditions. Most of the natural resources are directly or indirectly related to the surface cover in a given locality. Therefore, to maintain harmony among sustainable resources and socio-economic needs, land cover and land use studies should be dealt with care. Land use/land cover information at various levels forms a basic input for all planning and management activities of an area or region. Utilization of land has lead to all area of the earth being modified. The land system is highly dynamic which undergoes significant changes according to changes in socio-economic and natural environment. The changes in any forms of land use are largely related with the external forces and pressure built up within the system.

Assessment of spatial and temporal changes in land use pattern is an effective tool for the evaluation of changes occurring in land use and extent of environmental degradation. The great merits of recording land use/land cover on maps are that each parcel of land can be precisely located and the aerial relationship of different uses analysed in a way that is not possible with a variety of statistical data, often collected for quite different purposes. Remote sensing can be a powerful method for monitoring changes in land cover and land use.

Objectives

The aim of the present paper is to analyse land use/land cover mapping using satellite remote sensing digital techniques and to carry out change detection studies to study the temporal changes in the land use/land cover in selected study area.

The study has been carried out with the following objectives:

(i) To study and identify land use/land cover classification for the study area using the satellite remote sensing data and supervised classification algorithm.

(ii) To assess the change in land use/land cover in the study area between 2000 and 2005 AD through change detection studies.

Study Area

The study area selected for the present paper is Dhari taluka in Amreli district of Gujarat State. The taluka lies between 70°48′14.37″ and 71°16′4.77″E latitudes, and between 21°7′13.20″ and 21°28′21.01″ N longitudes.

Analysis of Land Use/Land Cover Mapping and Change Detection Studies

Remote Sensing Techniques for Land Use/Land Cover Mapping

Timely and reliable information base is essential for effective land use management. In the past, land use/land cover mapping was done by conventional ground methods, which are labour intensive, time consuming and carried out relatively infrequently. While much of the information can be derived from existing paper maps, which are not available for all parts of the country, these soon become outdated, particularly in a rapidly changing environment.

In recent years, satellite remote sensing techniques have been developed which are of immense value for preparing accurate land use/land cover map and monitoring changes at regular periodic intervals.

While land use/land cover information can be directly interpreted from appropriate remote sensing images through image classification and classification accuracy assessment.

Image Classification

Classification is one of the most often used methods for information extraction. This procedure assumes that the imagery of specific geographic area is collected in multiple regions of the electromagnetic spectrum and that the images are in good registration. The overall objectives of the image classification procedure are to automatically categorize all pixels in an image into land use/land cover classes or themes.

Most of information extraction technique depends on the spectral reflectance properties of various features covered in the scene and employ special algorithms designed to perform various types of spectral analysis. The process of multi-spectral classification can be performed using either supervised classification or unsupervised classification.

Change Detection Studies

Change detection involves the uses of multi-temporal data set to discriminate areas of land cover change between dates of imaging. The types of changes that might be of interest can be range from short-term phenomenon such as snow cover or flood water to long-term phenomenon such as urban fringe development or desertification. Ideally, change detection procedure should involve data acquired by the same (or similar) sensor and be recorded using the same spatial resolution, viewing geometry, spectral bands, radiometric resolution, and time of day.

Accurate spatial registration of various dates of imagery is also a requirement for effective change detection. Registration to within ¼ to ½ pixel is generally required. Clearly, when registration error is greater than one pixel, numerous errors will result when comparing the images. The reliability of the change detection process may also be strongly influenced by various environmental factors that might change between image dates.

Methodology

Land use/land cover mapping and change detection studies based on the digital processing of remotely sensed data need careful planning of various activities. Different land use/land cover categories were identified based on the image characteristics like tone, texture, size, shape, pattern, location, association, etc. Geo-database creation and statistics generation for land use/land cover and changes is done in GIS environment. The following activities, preparatory to digital processing are particularly relevant to successful transformation of remotely sensed data into land use/land cover categories.

Classification System

The current land use/land cover maps are prepared by the classification system given in Table 4.1. Any category of land use/land cover in study area, which is not covered under the preview of the classification system is put under the category "others".

Land Use/Land Cover Parameters

Change in area of different land use/land cover categories are obtainable directly from digitally classified land use/land cover map and change detection studies These changes in area with time, partially indicate the impact due to environment and human activities.

Data Used

Satellite data pertaining to two periods, i.e., IRS-1C/1D LISS-III and IRS-P6 LISS-III data of November 2000 and October 2005 are used for land use land cover mapping (Plate 4.1).

Table 4.1

Classification Scheme of Study Area

S. No.	*Level – I*	*Level – II*
1	Settlement	
2	Agriculture land	Crop land Fallow land
3	Forest	
4	Waste land	
5	Water bodies	
6	Others	

Geo-referencing of Satellite Data

The satellite data LISS-III IRS-P6 are geo-referenced to bring into the real world coordinates.

Preparation of Base Maps

Base maps are prepared by extracting base information from existing maps then updating from satellite images. The base layers such as drainage, settlements, road network, water bodies, etc., are necessary for identifying the suitable sites for the thematic layer of land use/land cover. After digital method is followed, digital output is to be converted into vector data for correspondence with other theme maps such as drainage (major rivers, stream and streamlets), water bodies, settlements, road and rail network have been drawn carefully.

Supervised Classification

The Parallelepiped method of supervised classification technique has been used for land use/land cover mapping and change detection studies.

Change Detection

Statistics of land use changes have been done in GIS environment. For change detection, the data of IRS-1C/1D LISS-III and IRS-P6 LISS -III for the year 2000 and 2005 are used.

The imagery has to be interpreted and checked for correction. After checking of collateral data the maps have been finalized.

Figure 4.1

Change in Area of Land Use/Land Cover between 2000 & 2005

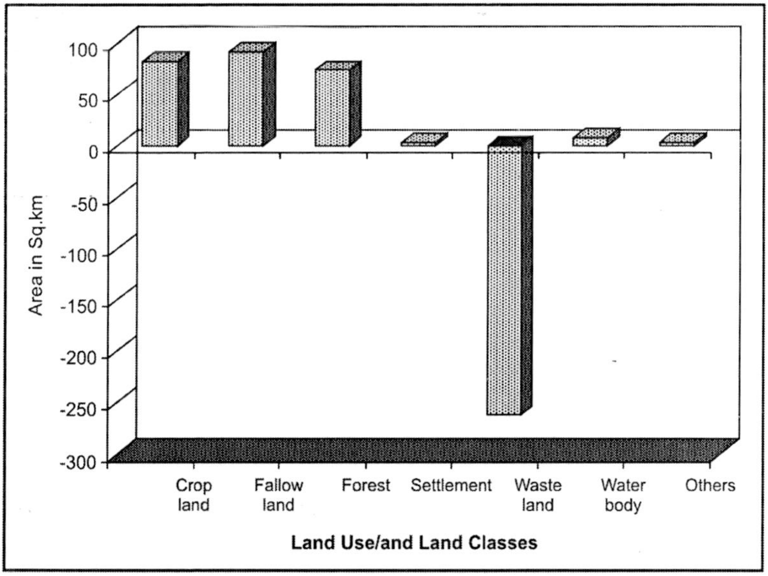

Result

Introduction

Land use and land cover mapping using remote sensing has been carried out for Dhari taluka. LISS III data set has been used and Parallelepiped algorithm has been used for digital classification. Change detection studies have been carried out in Dhari taluka to access the change in land use and land cover mapping and change detection studies for Dhari taluka are presented in this paper.

Land Use/Land Cover Statistics

Land use and land cover classes for Dhari taluka after digital classification is shown in Imgae 4.1 (page 60) for 2000 and 2005. Land use and land cover statistic are given in Table 4.2.

Change Detection Studies

Land use and land cover pattern in area has a dynamic character and is usually bound to change over time. The change in land use and land cover pattern may be due to human activities on the land or due to natural causes or may be due to human included natural causes. For proper management of any land and for generation of long-term development action plans it is essential to study the temporal changes of land use and land cover over certain fixed time interval.

Table 4.2

Change in Area of Land Use/Land Cover Class

S. No.	*Land use and land cover classes*	*Area (sq km)*	
		2000	*2005*
1	Crop land	116.95	199.38
2	Fallow land	463.10	555.03
3	Forest	79.27	152.95
4	Settlement	8.67	11.33
5	Waste land	370.84	109.44
6	Water body	0.80	7.87
7	Others	1.30	4.12

The change in the land use and land cover can be said to be positive if wasteland has been converted to good usable land, such as cropland or change can said to be positive if there is increase in the forest vegetation. On the other hand, the changes are said to be negative if it is vice versa, i.e., cropland has degraded to waste land or forest vegetation has got degraded.

Change detection has been carried out by analysis of multi-date LISS III data of November 2000 and October 2005 spanning almost five years (2000–2005). The results are tabulated in Table 4.3. It is also shown graphically in Figure 4.1

Table 4.3

Change in Area of Land use/Land Cover Class

S. No.	*Land use and land cover classes*	*Change in area (sq km) between 2000 & 2005*
1	Crop land	82.43
2	Fallow land	91.93
3	Forest	73.68
4	Settlement	2.66
5	Waste land	–261.44
6	Water body	7.07
7	Others	2.84

Conclusion

The objectives of this study were to interpret the land use level information and land cover features using satellite remote sensing digital techniques to generate land use and land cover maps for Dhari taluka. Apart from this change in the land use and land cover pattern between 2000 and 2005 has been accessed through change detection studies.

The study has shown the remote sensing techniques have tremendous potential for mapping of land use and land cover features of the region. Digital classification techniques have added advantage of better accuracy and faster output generation. GIS has proved to be a powerful tool for integrating these various information and generation of geo-database.

Total agriculture land in 2000 was 579.44 sq km and 754.41 sq km in 2005. In 2000 it was 463.10 sq km. It was shown as increased because few wastelands were converted in the agriculture land due to the fact that the year 2000 was a drought year and there was good rain in 2005.

In 2000, wasteland was 370.84 sq km but due to good rain in 2005 it was only 109.44 sq km. The change detection study shows that there is urban expansion by 2.66 sq km.

Forest area has increased by a large amount, i.e., 73.68 sq km. It is due to good rain between 2000 and 2005, and the human activity in favour of plantation.

Plate 4.1

Classified Image of 2000–05

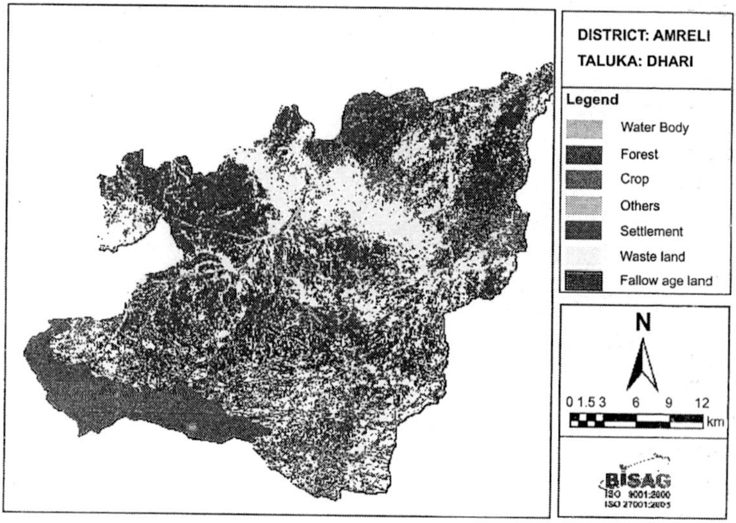

Plate 4.1(a) : Classified Image of 2000

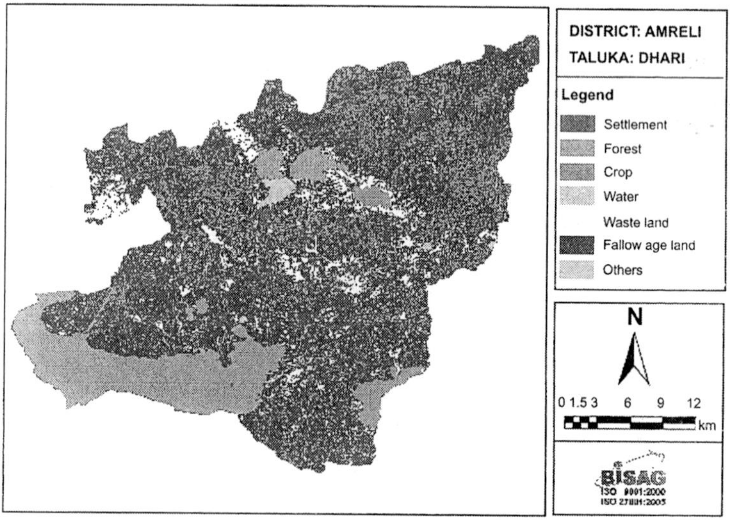

Plate 4.1(b) : Classified Image of 2005

Since 2000 was a drought year, large areas of forest was in the form of wasteland as evident in the image of 2000, which was converted into forest by 2005.

Water body area has increased by 7.07 sq km because 2000 was a drought year. So some water body region in 2005 was in the category of wasteland in 2000 as classified by supervised classification.

Acknowledgement

The author of paper is thankful to Mr Vijay Singh, Senior Scientist, BISAG, Gandhinagar for guidance during internship in the institute.

References

Anderson, J.R. (1971), 'Land Use/Land Cover Changes: A Framework for Monitoring', *Journal of Research*, Vol. 5, No. 3, U.S. Geological Survey, Government of India, pp. 143–53.

Gautam, N.C. and V. Raghavswamy (2003), *Land Use/Land Cover and Management Practices in India*, B.S. Publications, Hyderabad.

Keith, C. Clarke, Bradley O. Parks and Michael P. Crane (2006), *Geographic Information Systems and Environmental Modeling*, Prentice Hall of India Pvt. Ltd., New Delhi.

Lo, C.P. and Albert K.W. Yeung (2006), *Concepts and Techniques of Geographic Information System*, Prentice Hall of India Pvt. Ltd., New Delhi.

Reddy, M. Anji (2001), *Remote Sensing and Geographic Information System*, B.S. Publications, Hyderabad.

5

Groundwater Prospect Zoning by Remote Sensing and GIS Techniques in Dungarpur Block

Maya Poonia and Sunita Singh

Water is a precious natural resource in our world. It is essential for any kind of life on the earth. Out of the total water available on the earth, only three per cent is freshwater useful for human consumption. The availability of freshwater is decreasing continuously with time, due to the changes in the climate system, global warming, contamination of river and lake water, etc.

The demand for water has increased over the years and this has led to water scarcity in many parts of the world. The situation is aggravated by the problem of water pollution or contamination. India is heading towards a freshwater crisis mainly due to improper management of water resources and environmental degradation, which leads to a lack of access of safe water supply to millions of people. This freshwater crisis is already evident in many parts of India, varying in scale and intensity, depending mainly on the time of the year.

Rajasthan is the most water-deficient state of the country. It has 10.4 per cent of the country's area but only 1 per cent of its water resources in the form of annual rainfall. The topography is dominated by the Aravali range of mountains, which divides the state into two different sets of physiographic regions. The Trans-Aravali region is known as "Great Indian Thar Desert", while the Sis-Aravali region comprises of eastern plains and south-eastern Hadoti plateau. The desert occupies about 60 per cent area of the State with 40 per cent population and is characterized by extremely low and erratic rainfall. The eastern and south-eastern region covers only 40 per cent area with 60 per cent inhabitants. The unique geographical situation and distinct physiographic condition of the state is such that it receives the last residual share of monsoon rainfall from both the sides, viz., the Bay of Bengal and South-West monsoons.

A good part of it is absorbed in dry soils, thus restricting the surface runoff. Normally water absorbed in the zone of aeration should recharge the groundwater, but it does not reach up to the zone of saturation, thus leaving comparatively a small part of potable water that can be used for drinking or agriculture purpose.

The surface water resources in the state besides being scarce are mainly confined to south and south-east part of the State. The demand for water has increased over the years and this has led to water scarcity in many parts of the world. The situation is aggravated by the problem of water pollution or contamination. India is heading towards a freshwater end, mainly due to improper management of water resources and environmental degradation, which leads to a lack of access of safe water supply to millions of people. This fresh-water crisis is already evident in many parts of India, varying in scale and intensity depending mainly on the time of the year.

As a primary goal, National Natural Resource Management System (NNRMS) of Department of Space has initiated the studies of land and water resources for scientific planning coupled with agricultural, infrastructural, and regional development.

Objectives

Objectives of this paper are as follows:

(i) To discuss the problem of groundwater through analysis of thematic maps, i.e., – Base map, Drainage map, Geomorphology map, Lithology map, Structural map, Hydrology map and Groundwater prospect map.

(ii) To integrate thematic maps of Groundwater prospect map.

(iii) To examine patterns, drainage density, landforms and soil erosion in the study area.

Some of the advantages of this approach are: (i) image depicting the scene as though one is observing the area from a point at high altitude, and (ii) the display of spatial relation among surface features in the same context as in maps (George Joseph, 2004).

Digital Image Processing Technique

Digital image processing involves the manipulation and interpretation of images obtained in the digital form with the aid of computers. The digital data can be treated with various algorithms for enhancing the ground features for better interpretability and analysis. Various image-processing techniques are available for geometric correction, image enhancement, statistical analysis, arithmetic analysis, feature extraction and classification. Digital image processing system comprises hardware and software. The main advantage of extraction through digital image processing is the speed of operation and unbiased result.

Remote Sensing Applications in Groundwater Resource Evaluation

Considerable progress has been obtained in the last two decades in the application of satellite remote sensing techniques for groundwater resources evaluation and management. Applications such as groundwater prospecting using remote sensing techniques are now in operational mode.

Though groundwater is a natural resource which occur in subsurface, but many indirect surface indications of groundwater occurrence and movement can be mapped using remote sensing. These indicators include the lithology, structures, such as joints, fractures faults, lineaments, geomorphic units/land forms, hydrologic indicators such as surface drainage and water bodies, and land use/land cover.

The main advantage of remote sensing in groundwater resources studies are its synoptic viewing capabilities which helps in mapping regional-level geological structures, litho boundaries and geomorphic contacts, multi-spectral viewing which helps in discriminating between different litho units, landforms and land use/land cover units and repetitive coverage which helps in studying the same area in different seasons and different periods. The microwave remote sensing data has an added advantage of ground penetration which helps in bringing out subsurface feature of groundwater importance like buried/palaeochannels.

The use of remote sensing technique for groundwater studies helps in narrowing down considerably the areas for detailed groundwater exploration, thus avoiding the labourious expensive and time-consuming extensive field survey.

Study Area

The study area of Dungarpur district is located in the southern part of Rajasthan in between 23°20′ N, and 24°01′ North latitude, and between 73°21′ E and 74°23′ E longitude. In the west, it has common border with Gujarat. The total area of the district is 3,770 sq km. In the north and east, the landscape is rugged and wild, but towards the south-west border, the features seem to merge in the topography of Gujarat region. The eastern portion slopes down towards the basin of Mahi. The rugged and wild aspect of the region is attributed to the offshoots of the Aravallis. Two perennial rivers, viz., Mahi and Som flow through the district. The Mahi separates the district from Banswara, and Som forms a natural boundary with Udaipur district. The non-perennial streams are Jakham, Majhan, Vatrak, Bhader, Gangli, Sapan and Veri Ganga. The water table in the district varies from 5 to 15 m below ground level. The normal annual rainfall in the district is about 761.7 mm. The district has a dry climate with a hot season milder than in the desert region of Rajasthan. The maximum temperature in summer is about 38°C and minimum temperature in winter is about 9°C.

Geology

The geological system of the district belongs to the Pre-Cambrian Aravali system. Slates are in abundance in the central region and are largely identified with veins of quartz particularly in the area north and east of Dungarpur town. Some patches of pegmatite and granite intrusions may also be found in the slates. The ultra basic rocks have been observed as potential sources of asbestos, chromite, magnetite and talc. Among minerals, fluorite, beryl, asbestos and soapstone are the main ones. Besides, copper, mica, cyanide and lead ores have also been found.

Lithology

The lithological assemblage of the Dungarpur block forms southern continuity of Kherwara-Rikhabdeo area which belongs to the oldest known rocks of the Mangalwar complex and Aravali super group or rock, with intrusive ultramafic suite of rock. The generalized stratigraphy is given as under:

Mangalwar Complex

The oldest rocks of the Mangalwar complex observed, are weathered at the upper surface to give rise to a thin regality mantle. This complex is characterized by igneous rocks of several types and times of several types and times of intrusions, which have been intimately injected one with another, welding them into streaky and banded composite gneisses at many places. The individuality of original constituents are lost and the complex shows wide variability of texture and general lack of definition in its composition. The strike of foliation varies from N – S to NNW – SSE with generally steep dips with some prominent quartz veins in the area forming inliers, are exposed in the core of anticlinal valleys in the eastern and south-eastern parts of the block. These consist mainly of biotitic gneiss and schists, grannies gneiss and felspathised rock

Udaipur Groups

Overlying the gneissic rocks of the pre-Aravalis are the quartzites (Gurali-quartzites), forming high rising ridges in the area which are narrow crests. The quartzite is massive, silicified, well-jointed and depicts bedding traces having steep dips.

Overlying these quartzites as well as directly overlying the older gneissic rocks is a sequence of phyllites and schists (*nathara-ki-pal*) which very prominently exhibit a N – S folition trend along which are intrusive bodies measure from a few metres to more then one km in length. The phyllites-schists assemblage represents a flysch like coarse to fine-grained lithology having minor intercalations of chemogenic/biogenic rocks.

Ultramafics

The Rikhabdeo ultramafic suite of rocks can be traced in this block as a southern continuity into the river where it is cut out by a NNW trending major fault. This zone is widest in south-west of Manpur village and is characterized by its distinct darker tones. These rocks also occur as isolated patches in the north-western and northern parts of the area. This ultramafic suite of rocks consists of serpentine talc-carbonate rocks and chlorite schists, characterized by green colour

Quaternary

The Quaternary deposits consist of valley-fill and alluvium. The alluvium is confined to narrow zones along the banks of Som River and other larger streams. In rest of the mountain, most of the stream courses which generally follow narrow and pasture they are covered with valley-fill material of varying depths, consisting of coarse clastics.

Structure

The most significant structural feature of the area is the zone of the ultramafic suite of rocks which are considered to be emplaced along a deep-seated suture which is now represented by a lineament, having a general N – S trend. Its linearity also conforms to the main schistosity of the phyllites and schists assemblage into which has intruded this suite of rock also show three main sets of the phyllites and schists assemblage into which it has intruded. These rocks also show three main set of joints/fractures which give rise to a cuboidal nature to its otherwise schistose character. These mainly strike NE – SW, NEE – SSW and NW – SE. Their intersection points are considered to be prospective locales for groundwater survey.

A major semiformal structure within the Phyllis metagraywacke assemblage is mark between Bedsore and Obri, in Sagwara and its adjacent, a major flood closer is observed in the Dungarpur block around Nawagaon which has an E – W axial plane trace.

The disposition of the folds trends of the axial plane traces, foliation strikes and overall geometry suggest a multiphase folds affecting these metamorphic rocks, the most dominant and persistent fold structures, are nearly N – S trending which are a result of the second phase of folding, giving rise to sub-parallel foliation. The E – W to NE – SW trending folds are reacted to the third phase of NNE – SSW and have been obliterated by the later phases of folding. These can only be seen in the form of relict structure or by the intersection planes of planet structure.

Many lineaments have been marked in the area, most of which represent only major fractures in the varied lithological assemblage. However, some of this can be identified as faults responsible for disruption of geological contacts. One of these faults can be traced in the area south of Dungarpur which has NW – SE trend and which controls the row of Sandri River and part of Moran River and can be traced over a distance of 40 km in the Dungarpur and Sagwara blocks.

The terrain shows a highly dissected nature due to a large number of river segments; streams and channels show structural control along the joints.

Geomorphology

The district area is characterized by a mature topography dominantly reflecting denudational and structural landforms; the denudational include denudational hill pediments, pediplain, buried pediplain developed on different lithological units as demarcated from the interpretation of IRS, LISS-II FCC imagery of the months of April, October and June, 2000 supported by Survey of India toposheets and existing geological information.

The northern limit of the district is marked by the River Som which flows from west to east in a sinuous pattern. The south-eastern limit is marked by Mahi River which in this section flows from NNE – SSW in structurally aligned section The dam is made over Jakam river from the NE near Oda village, before finally joining Mahi river at Baneshm.

The highest elevation in the area is marked 449 m above MSL on the quartzite hill near Amihera in the extreme west of the district. Comparable heights are noted only in the western part of Dungarpur.

The drainage is entranced in nature and shows narrow linear valleys which are mainly structurally controlled with valley-fill deposits. Flood plains are observed only as narrow linear patches along the margins of Som, Moran and Sandri river courses. As the rivers have generally carved their courses in the hard rocks, the thickness of the flood plain deposits is not likely to be more than a couple of metres and is supported by rocky beds of these rivers.

Drainage and Watershed

The development of land and water resource on a sustained basis without deterioration and with constant increase in productivity is the main stay of mankind. Watershed or hydrologic units are natural hydrologic entities that cover a specific areal expanse of land surface from which the runoff flows to defined drain channel, stream or river at any particular point. These units are considered more efficient and appropriate for assessment of resources and subsequent planning and implementation of various development programmes. The watershed approach is more rational because land and water resources have optimum interaction and synergistic effect when developed on watershed basis.

The watershed approach is therefore, increasingly being employed in various development programmes like soil and water conservation, command area development, erosion control in catchments of river valley projects and flood-prone rivers, dryland/rained farming, reclamation of ravine lands, etc.

Dungarpur block forms the origin of some streams, viz., Som, Suri, Nori, Moran, Vatrak and Bhader and forms four major water divide. The River Som flows further east and then south to join Jakham and finally Mahi. The streams Nori and Moran flows east to meet the right bank of Mahi river. The Bhader river flows down south to meet Mahi river. Apart from this, the Vatrak stream which originates from the southern region of the block flows south-west to meet Hathmati and finally Sabarmati rivers.

Table 5.1

Details of Som Kamala Amba Dam

Complete	1728 Lacs Gamester (6102 MCFT)
Useful	1603 Lacs Gamester (5661 MCFT)
Irrigation Potential	Above 80%
Maximum Height of Chief Dam	620 metre
Maximum Height of Dam	34.50 metre
Flood Disposal Capacity	20500 Gamester Per Second
Length of Dam	2420 metre
Water Storage Area	3618 Hect.
Iron Gates	13 Nos.
Expected Life of Dam	100 years

There are three perennial rivers in the district namely Realm, Som and Magam. There is one medium irrigation project, 51 irrigation tanks and 900 village tanks to store rain water. In the district about 12,870 million cubic feet (MCFT) water can be stored out of which at present 3846 multi water is being stored in the irrigation tanks and 6,102 MCFT in the Som Kamala Amba.

The gross irrigated area of the district during 2000–01 was 27020 hectares while net irrigated area was 24,914 hectares (Table 5.1).

Table 5.2

Year-wise Irrigated Area by Different Sources

(Area in Hectares)

Year	*Well*	*Tanks*	*Canal*	*Other Resource*	*Irrigated Area*
1991–92	16109	7037	911	3022	27079
1992–93	16073	7659	1426	3400	28558
1993–94	17296	7841	1926	4052	31115
1994–95	17793	7840	1981	3906	31520
1995–96	17044	5490	3037	4250	29721
1996–97	20413	6817	4164	4034	35428
1997–98	22086	6298	4752	3317	36453
1998–99	21620	5444	5447	3833	36344
1999–2000	14236	1603	5684	2549	24072
2000–01	7618	478	3564	1570	13230
2001–02	13766	2744	8521	2479	27520

Slope

Slope is a conspicuous element of surface configuration and is a vital, determining parameter (of land facet in general and physical component in particular, reflecting its imprint on the overall flow pattern both surface and subsurface). It provides essential and inevitable input base for land irrigability and capability. Integration with slope perspective for sustainable development can help in understanding their closer interdependency with other resources.

The rugged and wild aspect of the region is attributed to the offshoots of the Aravalis – the country is broken and hilly, but none of the hills attains a great height. Large variation in slope categories is observed within the block. The north and north-eastern part of the tehsil is dominated by slopes less then 3 per cent with intermittent rise of hills here and there. Whereas, the central and southern cone of the block is dominated by higher slope categories (i.e., more than 15%) and this extends somewhat to the north, forming a north-south ridge. The area is very undulating with an intermixing of various slope categories occurring at close intervals. There are a few sudden rising hills in the moderately plain areas, and it generates a strongly sloping area.

The direct impact of slope is most vivid in the surface run-off manifested through drainage patterns, its densities, length and channel structuring which in turn determines: (a) the location of settlements, (b) location of embankments, anicuts, causeways, dams (both small and big), reservoirs, lakes, ponds and tanks, and (c) the size, shape, number and character of catchments and watersheds (sub, mini and micro).

Table 5.3

Area Under Various Slope Categories in Dungarpur Block

Slope Symbol	*Slope Category*	*Slope Percentage*	*Area in hect.*	*% of Total Geog. area*
1	Newly level	0–1	21,742.91	35.68
2	Very gently sloping	1–3	4,555.23	7.48
3	Gentle	3–5	2,286.30	3.75
4	Moderately sloping	5–10	9,229.38	14.97
5	Strongly sloping	10–15	5,781.06	9.49
6	Moderately steep slope	15–35	14,731.11	24.18
7	Very steep slope	> 35	1348.84	2.21
8	Water bodies/ streams/rivers		1366.16	2.24
Total			60,930.99	100.00

The slope map of Dungarpur shows all the seven categories of slope areas differentiated on the basis of percentage slope. Category-wise area and their respective percentage of the total geographical area are depicted in Table 5.3.

Tools and Techniques

Data Used

Slope Map, Geomorphological Map, Drainage and Structural Map are the thematic maps prepared for this study. Features on toposheets or other analog documents were scanned. They were imported and, geo-referenced then digitized by 'on screen digitization' in ERDAS Imagine. Toposheets (46 I/1, I/2, E/13, E/14), ETM FCC and published literatures have been used.

The ground truth verification is also done as and when required at various stages. The local knowledge of the area has been taken into consideration at the time of interpretation. All drainages are taken out from toposheets and further modified by satellite data. The groundwater prospect of the delineated geomorphic units are evaluated using available hydrologic characteristics and aquifer parameters.

Satellite Data

For better results Landsat's ETM image is used. Landsat current flagship sensor, the enhanced thematic mapper (ETM), was placed in orbit in April 1999, with the successful launch of Landsat 7. This system was designed by NASA and is operated by NOAA and USGS. ETM has also a 15 m panchromatic channel. Swath width remains at 185 km, and the system is characterized by improvements in accuracy of calibration, data transmission and other characteristics.

Table 5.4

Spectral Characteristics of ETM

S. No.	Band	Spectral Range (Micron Met.)	Ground Resolution
1	Blue	0.45–0. 52	30
2	Green	0.52–0.60	30
3	Red	0.63–0.69	30
4	Near Infrared	0.76–0.90	30
5	Middle Infrared 1	1.55–1.75	30
6	Thermal Infrared	10.40–12.50	60
7	Middle Infrared 2	2.08–2.35	30
8	Panchromatic	0.50–0.90	15

Topographic Data

Elevation data is used for various application, most notably the production of topographic maps and three-dimensional visualization of the earth's surface. Elevation data is used in analyses as a means both for explaining processes and for predicting them through modelling. The methodology involved the following procedures:

- Geo-referencing of toposheets and image
- Digitizing and editing
- Ground truth
- Thematic map preparation
- Assigning rank and weightage to various thematic layers
- Integration of thematic layers

General Methodology for Database Generation and Image Processing

Grid Base Generation

The grid base will be generated to register the electronically captured maps in the form of raster image, or vector format.

Data Loading

Both the year's data are loaded through import/export function of ERDAS Imagine. From the original data sets area of interest (AOI) is extracted and taken to image processing environment. Before preparing mosaic and further processing, data quality is thoroughly checked.

Georeferencing of ETM Data

Registration of the satellite data through Geometric Correction function of ERDAS Imagine.

Software Used

ERDAS Imagine 9.0

ERDAS Imagine is a standard image processing software. It has a high degree of administration flexibility, new space-saving features; many import and export utilities, greater mobility for fieldwork, superior mosaicking functions, and advanced 3D visualization and scene creation capabilities. It also includes numerous image enhancement techniques for the purpose of better image interpretation and extracting more reliable information.

ARC/INFO 9.1

Environmental System Research Institute's (ESRI) ARC/INFO Version 9.1 marks another step for professional GIS. Arc Info 9.1 enables user to improve spatial database construction, advanced visualization and analysis, advanced cartography, and spatial database management. Arc Info runs all applications and works with spatial databases. It is a more stable release and can provide a better platform for running your existing applications. This includes Arc Macro Language (AML) as well as Ordinary Differential Equation (ODE)-based applications. After completing digitizing, most of editing work is performed in Arc Info.

Other softwares used for study are Arc Map, Arc View 3.2, etc.

Implementation of Methodology

Assigned Rank and Weight to Different Thematic Layers

Mostly five thematic layers were used in this plan; these are drainage, geomorphology, slope, geology and structure. For further processing they were assigned rank and weight to thematic layers according to characteristics of the study area. Rank and weight of all layers are shown in a tabular form (Tables 5.5 to 5.9).

Table 5.5

Assigned Rank and Weight to Drainage Order

Drainage Order	*Rank*	*Weight*	*Rank × Weight*
1	5	1	5
2		2	5
3		8	5
4			30
5		7	35
6		8	40
7		9	45

Table 5.6

Assigned Rank and Weight to Lithology

Lithology Unit	*Rank*	*Weight*	*Rank × Weight*
Quartzite	2	1	2
Ultramafic Rocks		2	4
Gneiss		5	10
Phyllis		6	12

Table 5.7

Assigned Rank and Weight to Geomorphology

Geomorphic Unit	*Rank*	*Weight*	*Rank × Weight*
Structural Hills	4	1	4
Denudational Hills		2	8
Buried Peneplain		5	20
Flood Plain		9	36
Dry Water Body		7	28
Pediplain (Phyllitic Rocks)		6	24
River		8	32
Perennial River		8	32
Water Body		8	32
Valley-fill		9	36

Table 5.8

Assigned Rank & Weight to Slope Category

Slope Class	*Rank*	*Weight*	*Rank × Weight*
Very Steep Slope	3	1	3
Moderately Steep to Steep		3	9
Strongly Sloping		5	15
Moderately Sloping		6	18
Gently Sloping		7	21
Very Gently Sloping		8	24
Nearly Level		9	27

Table 5.9

Assigned Rank and Weight to Structure

Rank	*Weight*	*Rank × Weight*	
Lineament	1	8	8
Structure Type		9	9

Figure 5.1

Schematic Diagram Showing Methodologies

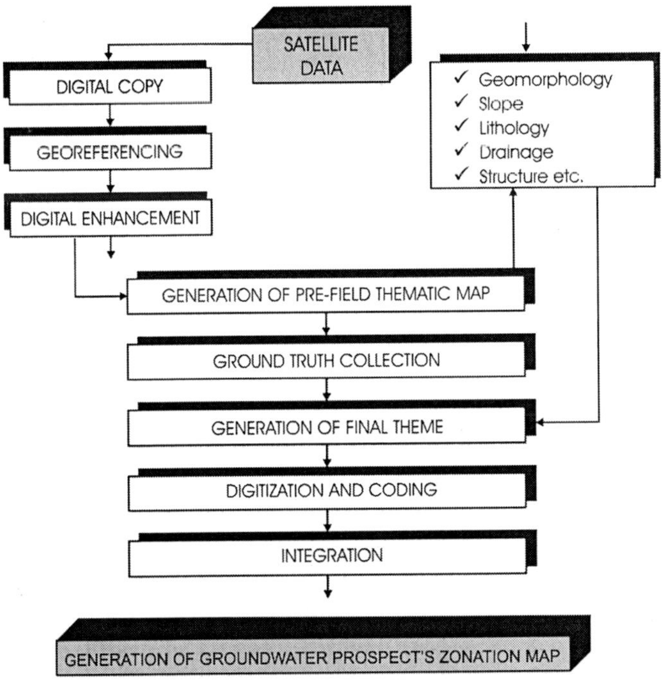

Figure 5.2

Schematic Diagram Showing Integration Steps of Raster Data

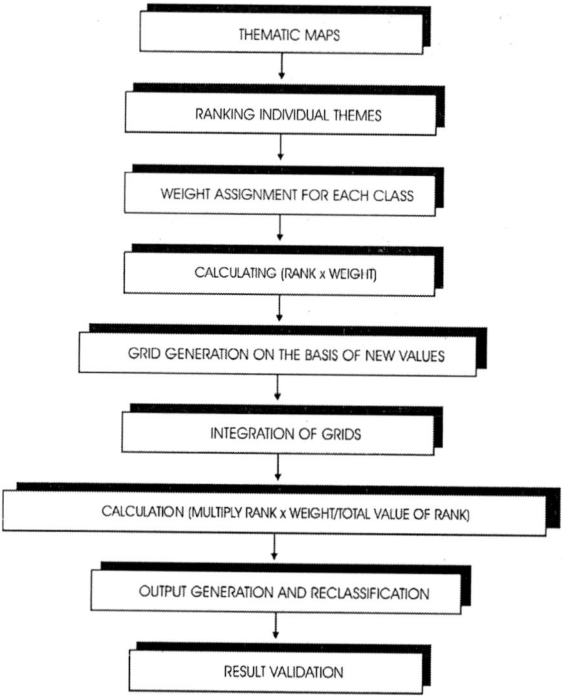

Conversion into Raster (Pixels) Data

Completion of the calculation, thematic layers have been converted into raster (pixel) format for further processing through feature to raster convert function in Arc Map.

Application of GIS Techniques for Integrated Maps

Integration of various thematic maps were performed by overlaying one theme map over the other, through cross operation of raster maps in Arc Map software. The cross operation performs an overlay of two raster maps by comparing pixels at the same position in both maps and keeping track of all the combinations that occurred between the values or classes in both maps. The geo-reference of input raster maps was kept the same during the cross operation; combinations of class names, identifiers or values of pixels in both maps were listed and the number of pixels occurring in this combination was computed. Rank and weight of all the thematic layers were multiplied by each other and divided by total of rank values. The results were stored in the output cross map and in a cross table. The output cross table obtains an identifier domain, with the same name as the output map. This domain contains items, which were combinations of the class names, IDs, groups, name or values of the first input map.

Results and Discussions

Detailed groundwater potential map has been prepared on the basis of hydrogeomorphology, surface characteristics, lithology, structure and secondary data on water level, water fluctuations, aquifer characteristics, etc. The map shows major water potential units on 1:50000 scale. The main hydrogeological units occurring in Dungarpur are granite gneisses of banded gneissic complex phyllites and schists and ultra basic rocks of Aravalis Super group. Dungarpur block mostly has phyllites and schists acquifers (366.74 sq km) followed by granite gneisses (56 sq km). Fluctuation ranges from 1.23 to 7.29 m in phyllites and schist zone and 1.85 to 2.84 m in granite gneisses acquifers with an average fluctuation of 3.96 m. In Dungarpur, undulating topography plays a dominant role in potential areas of groundwater recharge. Most of the rain water flows down along slopes as surface runoff towards the drains, ephemeral streams and rivers and a small amount of it is stored in the topographical depressions such as basins, valley-fills and flood plains. Because of the limited extent of such geomorphic units, the overall groundwater prospect is limited. The major hydrogeological zone is phyllite/schist with poor groundwater prospect zone and the depth at which the water is available ranges between 4–10 m. Overlaying the gneissic rocks of the Pre-Aravalis are the quartzite, forming high rising ridges in the area which have narrow crests. The quartzite is massive, silicified, well-jointed and depicts bedding traces having steep dips. The groundwater condition in valley-fills is moderate, in irrigated and flood plain; moderate to good, in buried pediplain area; moderate, to poor in structural and denudational hills. Most of the lineaments are confined to the valley regions and these lineaments are acting as pathways for groundwater movements.

A number of spot observations were made by covering almost all types of landforms slope, groundwater potential and quality, rainfall and climate, present land use, etc. The land characteristics as

Plate 5.1

Groundwater Potential Zone of Dungarpur Block

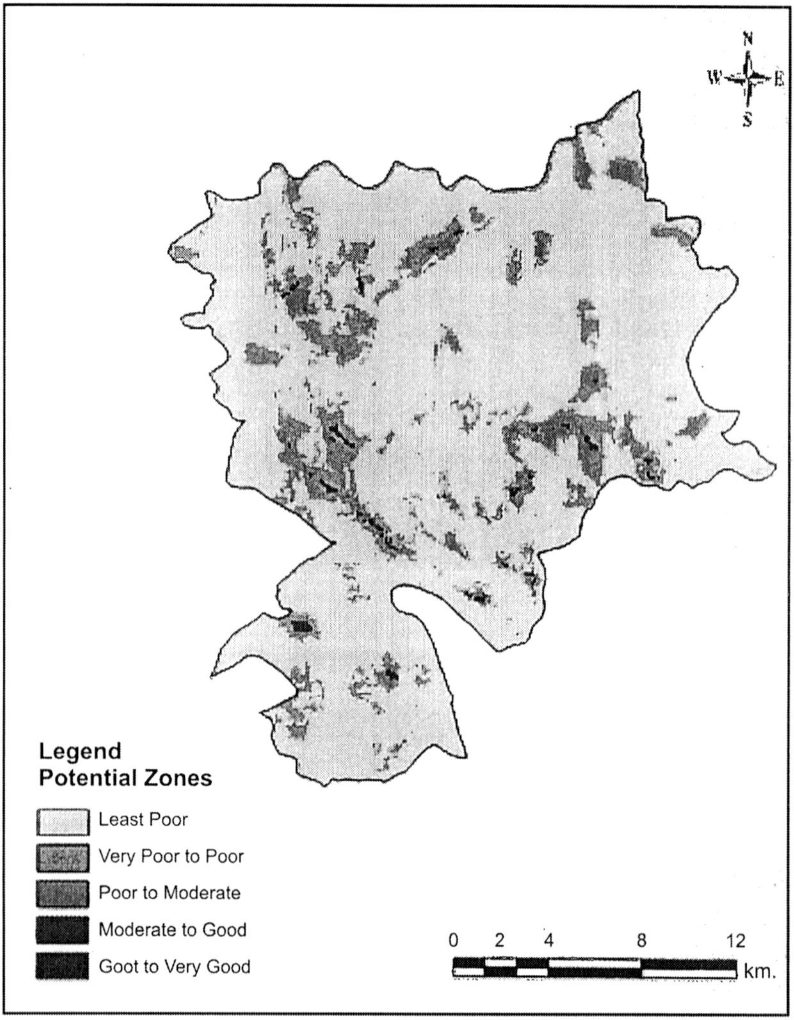

mapped in the respective theme maps were recorded along with the present land use. The cross map, of all themes shows all combinations, which were possible through integration procedure.

Conclusion

Remote sensing and GIS technology has proved to be very useful for groundwater prospecting and water resource planning activities where the potential gap is still to be filled up. It is concluded that this type of study helps in making efficient use of groundwater resources and their conservation for future use. It also helps in better land use planning and regional development;

The ETM imagery, is described on the basis of hydro-geomorphological interpretation coupled with Survey of India toposheets and existing geological information. The study also helps in the identification of potential groundwater zones. Table 5.1 gives the details of hydrogeomorphological units with their groundwater prospects.

Groundwater Prospect

The groundwater prospect in the block is moderate to good. The terrain is essentially formed of phyllites/schistose rocks, gneisses and ultramafics, which do not form very good aquifer in general, except in few places where secondary porosity (faults, fractures intersections of joints, foliation and cleavages) permits percolation of rainwater. There is also very thin soil cover on this lithology, except in flood plains and valley-fills areas where depth of Quaternary cover is only sufficient to warrant shallow to medium depth wells and low duty tubewells.

The quality of groundwater is generally suitable for irrigation and drinking purposes except in a few localities where fluoride is high. The yield of dug wells range from 30,000–45000 lpd in phyllites/schistose 30,000–40,000 lpd in gneisses and ultramafic.

Acknowledgement

The authors of the paper are thankful to Dr K.R. Khichar, Sr. Scientists of Remote Sensing Division, B.M. Birla Science and Technology Centre, Jaipur for guidance during internship in the institute.

References

Anderson, J.R. (1971), 'Land Use/Land Cover Changes: A Framework for Monitoring', *Journal of Research*, Vol. 5, No. 3, U.S. Geological Survey, Government of India, pp. 143–53.

Gautam, N.C. and V. Raghavswamy (2003), *Land Use/Land Cover and Management Practices in India*, B.S. Publications, Hyderabad.

Jensen, J.R. (1975), *Digital Image Processing*, Prentice-Hall, Englewood Cliffs, New Jersey.

—— (2004), *Remote Sensing of the Environment: An Earth Resource Perspective*, University of South Carolina, Columbia, USA.

Joseph, G. (2004), *Fundamentals of Remote Sensing*, Cambridge University Press, Cambridge, UK.

Keith, C. Clarke, Bradley O. Parks and Michael P. Crane (2006), *Geographic Information Systems and Environmental Modeling*, Prentice Hall of India Pvt. Ltd., New Delhi.

Reddy, M. Anji (2001), *Remote Sensing and Geographic Information System*, B.S. Publications, Hyderabad.

6

Natural Resources Analysis by Using Remote Sensing and GIS Techniques

A Case Study of Papum Pare District, Arunachal Pradesh

Akshay Manohar and Yuv Raj Singh Rathore

Resource management is a very important task and requires the combined efforts of different experts, ranging from environmentalists to economists and administrators. The Arunachal Pradesh is one among the seven north-eastern states of India which has rich natural resources. Due to the non-existence of natural resources database, the development of the state is hampered to a great extent. The Government of Arunachal Pradesh realized the need to develop a natural resources database for the use of resource managers and planners. With the availability of high resolution satellite data, the remotely sensed data has become an obvious choice for the scientific community for natural resource mapping. The advancement in Geographic Information System (GIS) facilitates the storage, management and manipulation of the spatial as well as the non-spatial attribute data efficiently.

Natural resources particularly land and water resources play an important role in the development of any region. The inventory of location and status of these resources are vital in development planning. Conventional ground-based surveys to acquire the natural resources information are time consuming and expensive. Advent of Remote Sensing technology has paved ways for preparing the base-line data of these resources on time and cost-effective mode. Remote sensing technology coupled with the GIS is a proven approach for preparing the resource base. Especially, these techniques would be of immense help in the mountainous and inaccessible terrain like in Arunachal Pradesh.

Objectives

The objectives of the research paper are as follows:

- To prepare thematic maps by using satellite image interpretation and group truth, such as land use/land cover, slope and aspect, hydrogeomorphology, forest type and density, and soil resources.
- To create of digital database of the above thematic maps in Geographic Information System (GIS) environment.
- To prepare Natural Resources Database maps for the area under study.

Study Area

The name Arunachal Pradesh meaning "The land of rising sun" was given to this state as it is located in the extreme eastern part of India and receives the first rays of the rising sun. The state of Arunachal Pradesh covers a geographical area of 83,743 sq kms, which accounts for about 2.55 per cent of the total area of the country. Arunachal Pradesh is the largest state (areawise) in the North-East region. It lies between the latitudes 26°28′N to 29°30′N and the longitudes 91°30′E to 97°30E.

Administrative Set-up

Itanagar is the capital of Arunachal Pradesh and is located at an altitude of 530 metres above mean sea level MSL in Papum Pare district. Itanagar is named after Ita Fort meaning fort of bricks built in 14th century AD. Presently, there are 16 districts in the state. Details of administrative set-up are given in Table 6.1.

Table 6.1

District of Arunachal Pradesh

S. No.	*Districts*	*District Headquarters*	*S. No.*	*Districts*	*District Headquarters*
1	Tawang	Tawang	9	East Siang	Pasighat
2	West Kameng	Bomdila	10	Upper Siang	Yingkiong

Cont'd...

...Cont'd

3	East Kameng	Seppa	11	Dibang Valley	Anini
4	Papum Pare	Yupia	12	Lower Dibang Valley	Roing
5	Lower Subansiri	Ziro	13	Anjaw	Hawai
6	Krrung Kumey	Koloriang	14	Lohit	Tezu
7	Upper Subansiri	Daporijo	15	Changlang	Changlang
8	West Kameng	Along	16	Tirap	Khonsa

Source: Census of India 2001

Papum Pare District

District name is derived from the rivers Papum and Pare. The district is situated between 93°13′ to 94°13′ East longitudes and 26°55′ to 27°38′ North latitudes. The geographical area of the district is 3,427 sq km. The district is bounded by Kurung Kumey district in the north, lower Subansiri district in the north-east, Assam state in the south-east and south and east Kameng district in the east. The district was carved out from Lower Subansiri district on 22 September, 1992.

Administrative Set-up

Papum Pare district is divided into two sub-divisions namely Itanagar and Sagalee and 8 Circles. The district headquarters is Yupia, which is about 18 kms from the state capital, Itanagar. Itanagar Capital Complex forms part of this district and because of it, the district is benefited in terms of development.

Methodology

Thematic maps of land use/land cover, slope, hydrogeomorphology, forest type, forest density and soil resources have been prepared through visual interpretation of satellite imagery followed by ground truth collection. Available published maps and other ancillary information have been utilized as reference for finalization of thematic maps.

Land Use/Land Cover

Geocoded false colour composite (FCC) of the satellite image was visually interpreted and based on image characteristics, land use/land cover classes have been delineated. Doubtful areas have been identified for field checks. During the fieldwork, the satellite imagery interpretation has been validated and changes, where noticed, were recorded. Photographs of representative land use/land cover classes have been taken and their exact location is marked with the help of handheld GPS. Doubtful land use/land cover class type and boundaries have been checked in the field and the maps suitably modified.

Table 6.2

Land Use/Land Cover Classification

Land Use/Land Cover Category	*Land Use/Land Cover Class*
1 Built-up land	1.1 Village
	1.2 Town
2 Agricultural land	2.1 Settled cultivation
	2.2 Plantations/Orchards
3 Forests	3.1 Evergreen/semi evergreen forest (dense)
	3.2 Evergreen/semi-evergreen forest (open)
	3.3 Degraded/scrub forest
	3.4 Forest plantations
4 Wastelands	4.1 Land with scrub
	4.2 Land without scrub
	4.3 Barren rocky/Stony waste/Sheet-rock area
5 Water Bodies	5.1 River
	5.2 River sand
	5.3 Lake/Pond/Tank/Reservoir
6 Others	6.1 Current shifting cultivation
	6.2 Abandoned shifting cultivation
	6.3 Grassland
	6.4 Alpine grass
	6.5 Snow covered/glacial area
	6.6 Landslide/Open areas

Description of Land Use/Land Cover Classes

1. Built-up Land: It is an area of human habitation developed due to non-agricultural use. All man-made constructions, covering the land surface come in this category. These are human settlements comprising residential areas, transportation and communication lines, industrial and commercial complexes, utilities and services, etc. Collectively, cities and towns (urban) and villages (rural) are included under this category. Built-up land is categorized into towns and villages, based on 2001 census data.
2. Agricultural Land: It is defined as the land primarily used for farming and for production of food grains, fibre and other commercial and horticultural crops. It includes cropland (irrigated and unirrigated), fallow, plantations, aquaculture, etc.
   (i) *Settled Cultivation*: It is a land used for cultivation of crops. Settled cultivation includes those lands with standing crop as on the date of the satellite imagery.

(ii) *Plantations/Orchards:* It is an area under agricultural tree crops planted, adopting certain agricultural management techniques.

3. Forest: It is an area bearing an association predominantly of trees and other vegetation types capable of producing timber and other forest products.
   (i) *Evergreen/Semi-evergreen Forest:* A forest, which comprises of thick canopy of tall trees and predominantly remains green throughout the year, is called evergreen forest. It includes both coniferous and tropical broad leaf evergreen trees. Semi-evergreen is a mixture of both deciduous and evergreen trees, but the later is predominating.
      (a) *Dense Forest*: A forest area where the vegetative crown density is more than 40 per cent of the canopy cover comes under dense forest.
      (b) *Open Forest*: A forest area where the vegetative crown density is between 10 per cent and 40 per cent of the canopy cover comes under open forest.
   (ii) *Degraded/Scrub Forest:* This is a forest area where scrub is dominant. Scrub is a stunted tree or bush.
   (iii) *Forest Plantations:* It is an area of trees of species of forestry importance and raised in forest area. These forest areas are artificially planted with trees, either in the open spaces or by clearing already existing forests of economically inferior species.

4. Wastelands: These are degraded lands which can be brought under vegetative cover, with reasonable effort and which are currently underutilized, and lands which are deteriorating due to lack of appropriate water and soil management or on account of natural causes. Barren rocky/stony waste/sheet-rock areas category were only observed in my study area.
   (i) *Land with scrub:* This land is generally prone to deterioration due to erosion and covered with scrubs.
   (ii) *Land without scrub:* This land also prone to deterioration due to erosion and lack of vegetative cover.
   (iii) *Barren rocky/Stony waste/Sheet-rock area:* These areas are defined as rock exposures of varying lithology often barren and devoid of soil cover and vegetation.

5. Water bodies: This class comprises areas of surface water, either impounded in the form of tanks, ponds, lakes and reservoirs or

flowing as streams, rivers, canals, etc. Cooling pond, cooling reservoir, abandoned quarries with water bay also comes under this category, but they are not present in the study area.

(i) *River:* It is a natural course of water flowing upon the land surface along a definite channel.

(ii) *River sand:* It is the stabilized accumulation of sand in river.

(iii) *Lake/Pond:* A lake is a large body of surface impounded water, natural or artificial within the landmass.

Reservoir: A reservoir is an artificial lake created by construction of a dam across the river specifically for the generation of hydroelectric power or irrigation or water supply for domestic/industrial uses or flood control or in combination of these.

Tank: Tanks are small lakes or impounded water.

6. Others: All other land use/land cover conditions not included in any of the classes described earlier and those which are either area specific or with limited areal extent are included under this category.

(i) *Shifting Cultivation:* It is the result of cyclic land use practice of felling of trees and burning of forest areas for growing crops on forested hill slopes. It is locally known as Jhuming. In this system the cultivators clear the forest, burn the cleared vegetation and cultivate the land until the yield of crop falls below subsistence levels. At this point, the cultivators abandon the land and move to a new site.

Current shifting cultivation: As the name denotes these are the areas where they are under cultivation and left waste for less than one year after cultivation.

Abandoned shifting cultivation: These are the shifting cultivation areas, which are put to cultivation and left waste for more than one year but less than five years.

(ii) *Grassland:* This is a land with pastures/grass.

(iii) *Alpine grass*: This is grassland in which alpine type of grass is grown in the higher reaches in association with snow cover.

(iv) *Snow covered area/Glacial area:* Snow cover is defined as solid form of water consisting of minute particles of ice, whereas glacier is a mass of accumulated snow, and ice deforms under its own weight and starts flowing from the snow-covered areas.

(v) *Landslide/Open areas:* In areas of steep slopes with unconsolidated sediments mass, wasting will takes place in the rainy season. This is called landslide.

Digital Database

Individual land use/land cover maps prepared on sheets were finalized after incorporating changes observed in the field and each sheet is edge matched with adjacent sheet before they are cleared for quality check followed by digitization. These maps were then scanned in grey scale at 150 dots per inch (dpi) resolution and geo-referenced using ERDAS Imagine image processing software.

Figure 6.1

Flow Chart for Land Use/Land Cover Mapping

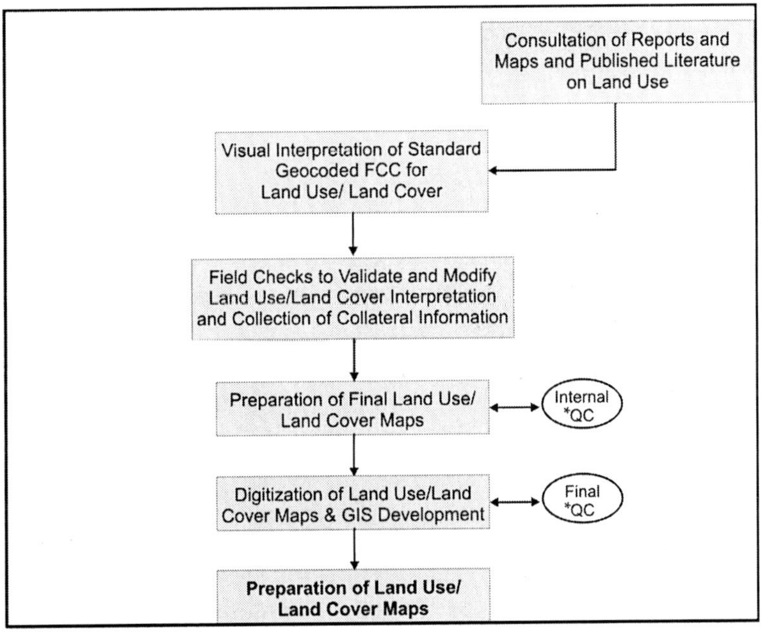

*QC: Quality Control

From land use/land cover maps, polygonal features of different land use/land cover classes have been digitized using AutoCAD software, each polygon having a distinct label of the class to which it belongs. AutoCAD drawing files have been converted in to Arc-Info softwares. Topology was built between different polygons and the overshoots and undershoots were edited. District-wise mosaics have been prepared after edge matching of the features between the adjacent sheets.

Detailed methodology followed for the preparation of land use/land cover maps is given in Figure 6.1.

Forest Type and Density

The Geomatics technology is used to identify the forest types and canopy density using satellite remote sensing data and other collateral data in the complex GIS environment. The multi-spectral, temporal, spatial and radiometric resolutions helps to assess and periodically monitor the status of forests and to assess the trend in the development of vegetation cover due to the impact of implementation of various management practices; to establish baseline data on the species compositions, sizes, densities, and distribution patterns within the various plant assemblages; and also to explore the structuring of the vegetation communities.

Vegetation Canopy Density Mapping: The multi-band IRS-1D, LISS-III satellite imagery in soft copy form is used for deriving the canopy density classes. Hybrid classification system is used. The imagery is segmented into homogeneous strata, using NDVI, supervised and unsupervised classification algorithms, using various image elements for interpretation of hilly and plain regions separately. The outputs are then compared and fused to derive the composite density map. The following classification scheme is used in line with Forest Survey of India.

Table 6.3

Forest Density Classification

Forest Density Class	*Description*
Closed Forest	All lands, with a forest cover with canopy density of 70 per cent and above.
Dense Forest	All lands, with a forest cover with canopy density of 70 to 40 per cent.
Open Forest	All lands, with forest cover with canopy density of 40 to 10 per cent.
Scrub Forest	All lands, generally in and around forest areas, having bushes and or poor tree growth, chiefly small or stunted trees with canopy density less than 10 per cent.
Non-forest Area	Areas without any forest cover.

Vegetation Type Mapping

- Expert classification system is used.
- The satellite imagery is segmented into various homogeneous strata using principal component analysis and canonical component analysis. Principal components analysis (PCA) is used as an exploratory tool to determine in what ways the plant assemblages were structured. The digital elevation model is also used as another variable to discriminate dominant community types.
- These strata are converted into a Forest Type Map with the help of existing working plans and reference material and other thematic layers like slope, soil, geology and geomorphology etc., along with inventory data available with the state.

These forest type maps are taken to the field for validation, and especially those on the fringes of transition of forest type were visited for ground truth verification, using the GPS. Along with ground truth, ancillary information on soil, climate, past treatment etc., were collected as this information also forms the basis of forest type classification. The ground truth information was integrated to arrive at the final forest type map.

Detailed methodology followed for the preparation of forest type and density maps is given in the following flow chart (Figure 6.2).

Figure 6.2

Flow Chart for Forest Type and Density Mapping

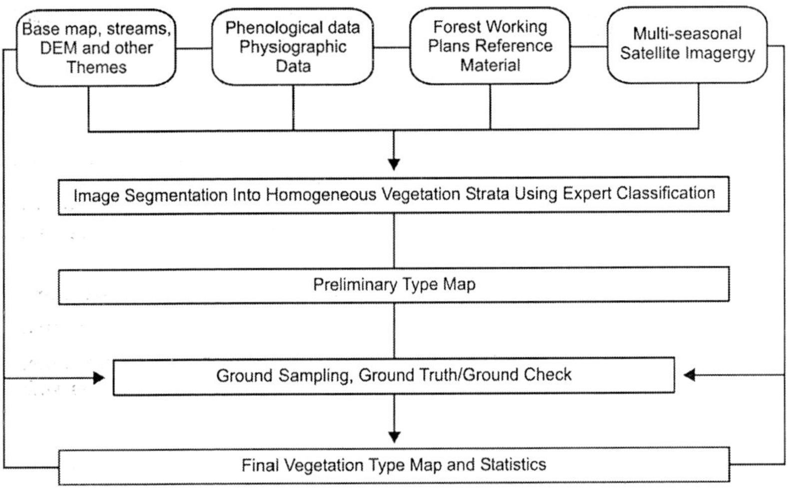

Forest type classification following for the preparation of forest type maps is given in the Table 6.4.

The datasets which are used for vegetation mapping are:

- Multi-seasonal and multi-spectral satellite imagery.
- Forest density and other thematic maps like the physiographic data, the topographical complexity, the drainage network, etc.
- Ground truth.
- Forest Working Plans and other collateral information.
- Other meteorological information.

Table 6.4

Forest Type Classification

S. No.	*Vegetation Cover Types*	*Forest Type as per Champion and Seth (1968)*
1	Dipterocarpus Forest	Assam Valley Tropical Wet Evergreen Forest (Dipterocarpus) (1B/C1)
2	Tropical Evergreen Forest	Upper Assam Valley Tropical Evergreen Forests (1B/C2)
3	Bamboo brakes	Cane Brakes (1B/C3/E1)
4	Tropical Semi-evergreen forest	Assam Valley Semi-Evergreen Forest (2B/C1)
5	Moist Deciduous Forest	East Himalayan Moist Mixed Deciduous Forests (3C/C3/3a)
6	Terminalia Forest	Eastern Hollock Forest (Terminalia myriocarpa) (3C/E1 (1S2a, b))
7	Grasslands	Eastern Wet Alluvial grassland (4D/2S2)
8	Riverine Forest	Tropical Riparian Fringing Forest (4E)
9	Subtropical Evergreen Forest	East Himalayan Subtropical Wet Hill Forests (8B/C1)
10	Pine Forest	Assam Subtropical Pine Forests (9/C2)
11	Temperate Board – Leaved Forest	East Himalayan Wet Temperate Forests (11B/C1)
12	Temperate Coniferous Forest	East Himalayan Mixed Coniferous Forests (12/C3a)
13	Temperate Scrub Forest	East Himalayan Sub-Alpine Birch/fir Forest (14/C2)
14	Rhododendron Scrub Forest	Birch Rhododendron Scrub Forest (15/C1)
15	Alpine Pastures	Alpine Pastures (15/C3)
16	Others	

Slope and Aspect

Slope is a measurement of gradient of steepness between two places or points on land surface. Slope is measured as ratio of vertical interval and horizontal distance of a given surface. The slope is expressed in percentages or degrees. Slope maps are prepared using 1:50,000 scale toposheets and the slope percentages are derived from the formula:

$$Slope\ Percentage = \frac{Contour\ interval\ in\ metres}{Distance\ between\ the\ contours\ in\ metres \times Scale\ factor} \times 100$$

State Remote Sensing Application Centre, Itanagar had provided the list of slope categories to be mapped. The categories are based on All India Soil and Land Use Survey (AIS & LUS) classification, modified to suit to north-eastern Himalayan terrain. The slope categories have been derived from contour spacing of topographical maps on 1:50,000 scale. The topo sheets are with contour intervals of 40 m and 20 m having been surveyed in different years.

The slope classification adopted in the present study is given in the Table 6.5.

Table 6.5

Slope Classification

S. No.	*Slope %*	*Slope Category*
1	0–5	Nearly plain to very gentle
2	5–15	Very gentle
3	10–25	Gentle
4	25–35	Gentle to moderate
5	35–50	Moderate
6	50–75	Steep
7	> 75	Very steep

Preparation of Digital Database

Individual slope and aspect maps prepared on Mylar sheets were finalized after edge matching with adjacent sheets before they are cleared for quality check followed by digitization. These finalized maps were scanned in grey scale at 150 dpi resolution and geo-referenced using ERDAS Imagine image processing software.

In the slope and aspect maps the slope categories are depicted as polygons and the aspect is shown as point feature, indicating any one of the eight compass directions. The slope category polygons are then digitized using AutoCAD software and converted into Arc-Info softwares. Topology was built between different polygons and the overshoots and undershoots have been edited and slope and aspect maps have been prepared. Aspect is depicted as a point symbol, showing the compass direction through an arrow symbol.

Hydrogeomorphology

Preparation of Geomorphological Maps

The mountains in Arunachal Pradesh owe their existence to the compressional forces associated with the movement of Indian plate towards north and north-east, signifying that the endogenic process (structure and tectonics) played a vital role in the origin of the landforms in the region. Accordingly, the hill ranges have been termed as structural hills, which are further classified into highly dissected, moderately dissected and least dissected structural hills, depending upon the degree of dissection. Other geomorphic processes operating upon the region are fluvial and denudational. In the higher reaches, the northern and north-east part of the state, glacial landforms have developed. Dominant processes and the land forms identified are given in Table 6.6.

Table 6.6

Geomorphic Units

Origin	*Land form*
Structural/Denudational	Piedmont Zone
	Pediment
	Valley
	Intermontane Valley
	Residual Hill
	Structural Hills
	Structural Hills – moderately dissected
	Structural Hills – highly dissected
Fluvial	Flood plain
	Older flood plain
	Palaeo channel
	Channel bar
	Stabilized channel bar
	Point bar
	River terrace
Glacial	Glacial Valley
	Terminal moraine
	Lateral moraine
	Snow covered hills

Significant geomorphic features identified are two levels of terraces at most places, mostly paired and local levels of terraces, and two major intermontane valleys, Ziro and Tale. While Ziro valley is used extensively for cultivation and human habitation, Tale valley is still ecologically undisturbed.

Groundwater prospects have been identified by integrating the geological and geomorphological maps in a GIS environment and assigning different prospect classes for each litho-geomorphic unit based on nature of material – consolidated/unconsolidated, porosity and permeability and secondary porosity due to intensive fracturing.

Description of Geomorphic Units

(1) Glacial origin

- (i) *Terminal moraine:* Glacial depositional landform, formed at the end of an ice stream.
- (ii) *Lateral moraine:* Glacial depositional landform, formed along the sides of an ice stream.
- (iii) *Glacial valley:* Valley in glaciated mountain area.
- (iv) *Glacier:* A river made of ice.
- (v) *Snow covered area:* Seasonal or permanent snow-cover area.

(2) Fluvial origin

- (i) *Flood plain:* A flat surface adjacent to a river composed of unconsolidated fluvial sediments. Normally, subject to periodic flooding by parent stream/river.
- (ii) *Older flood plain:* An area subjected to periodic flooding by the stream/river. Older refers to earlier cycle of deposition.
- (iii) *Palaeo channel:* An earlier river course filled with fluvial deposits of varying grain size and lithology, which is cut off from the main river.
- (iv) *Channel bar:* Sand bar formed in the braided river course due to vertical accretion of the sediments.
- (v) *Point bar:* Sand bar formed at the convex side of meandering river by lateral accretion of sediment.
- (vi) *Stabilized channel bar:* A channel bar, which is stabilized by growing of vegetation.
- (vii) *River terrace:* Flat upland adjoining the river course, occurring at different levels and comprises of gravel, sand and silt of river-borne deposits. It indicates the former valley floor.

(3) Structural/Denudational origin
   (i) *Valley:* Low lying depressions and negative landforms within the hills associated with stream/nala courses.
   (ii) *Intermontane Valley:* Broad depressions between mountains.
   (iii) *Piedmont zone:* Formed by coalescence of several alluvial fans by streams converging large area at the foothills.
   (iv) *Pediment:* Gently sloping, smooth surface of erosional bedrock between hill and plain with veneer of detritus.
   (v) *Residual hill:* Isolated low relief formed due to differential erosion so that a more resistant formation stands as residue, like small hills.
   (vi) *Structural hill:* Linear to arcuate hills showing definite structural trends. These hills have not been frequently dissected.
   (vii) *Structural hill highly dissected:* Linear to arcuate hills showing definite structural trends. These hills have been more frequently dissected by deep valleys/gullies.
   (viii) *Structural hill moderately dissected:* Linear to arcuate hills showing definite structural trends and dissected by deep valleys/gullies changing the original form considerably.

Preparation of Digital Database

Geological and geomorphological maps prepared on Mylar sheets were finalized after incorporating changes observed in the field and each sheet is edge matched with adjacent sheet before they are cleared for quality check followed by digitization. The maps were then scanned in grey scale at 150 dpi resolution and geo-referenced using ERDAS Imagine image processing software.

The geological map contains polygons (litho boundaries), lines (faults, lineaments, fold axis, thrust, etc.) and points (strike and dip symbols), which have been digitized with AutoCAD software and converted into coverages using ArcInfo software. Geomorphological maps have only polygonal features, which have been digitized and converted to coverages using ArcInfo software.

Results

Land Use/Land Cover

Dense evergreen/semi-evergreen forest cover account for about 80 per cent of the district. It is the dominant cover type in all the eight circles of the district but its areal extent is maximum in Balijan (620.53 sq km), Kimin (528.95 sq km) and Sagalee (609.29 sq km) circles.

Plate 6.1

Natural Resources Atlas of Arunchal Pradesh, Papum Pare District

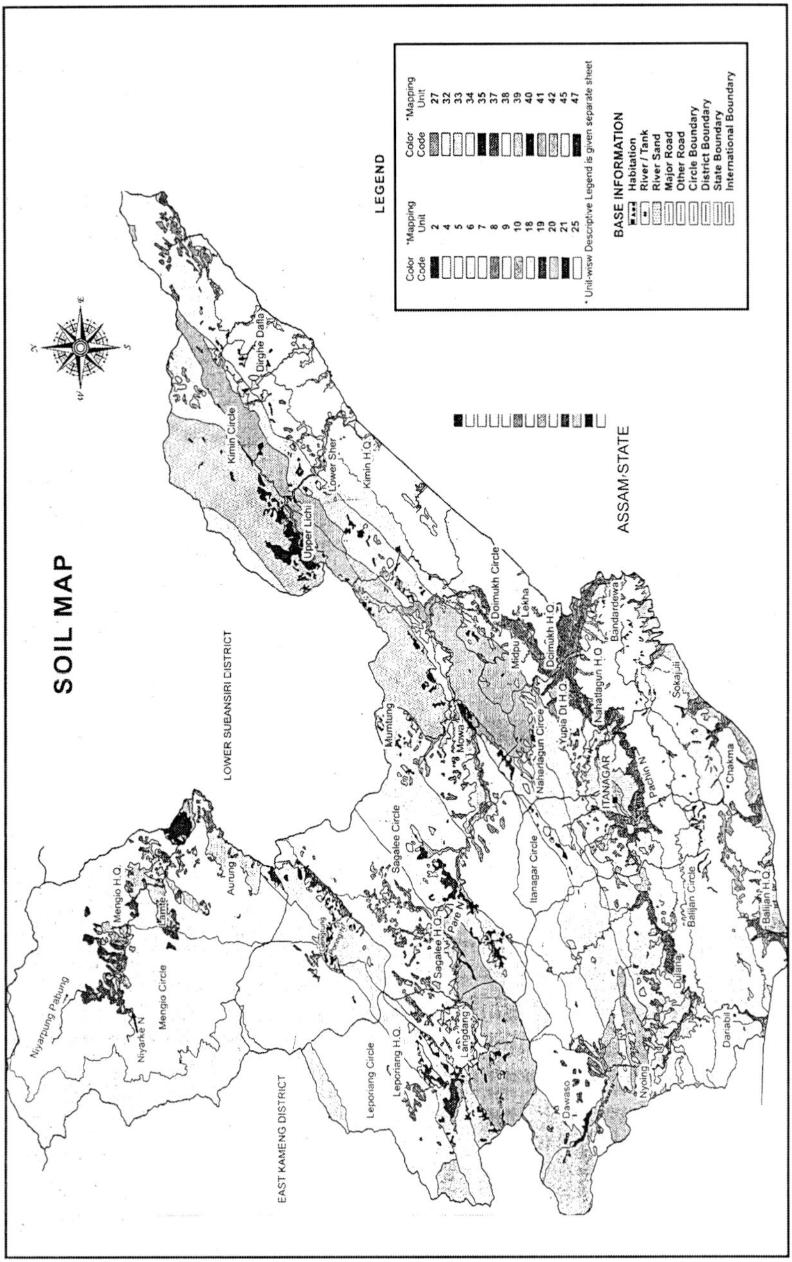

Degraded/scrub forest is the next extensive cover type well distributed in all the circles but maximum areal extent is noticed in Mengio (66.92 sq km) Baijan (51.24), and Sagalee (77.66 sq km) areas. Very limited extent of snow covered area is found in Mengio circle. It is surrounded by land with scrub, covering about 86 sq km. Settled cultivation along the floodplains and the river terraces is noticed in all the circles, but maximum areal extent is in Balijan circle (54.52 sq km) followed by Doimukh circle (13.40 sq km) and Naharlagun circle (13.0 sq km). Shifting cultivation (both current and abandoned) accounts for only 2 per cent of the district and is seen in all the circles, though the areal extent is maximum in Sagalee circle (19.25 sq km) followed by Balijan (10.02 sq km) and Naharlagun (9.49 sq km) circles. The areal extent of the towns is 22.34 sq km.

Conclusion

Both geological and geomorphological maps have been integrated in a GIS environment and a combined litho-geomorphic map has been prepared. Very small litho-geomorphic unit have been deleted. Groundwater prospects for the litho-geomorphic polygons were assessed based on lithological composition and geomorphology and categorized into very good, good, moderate, poor and very poor classes. Remote sensing and GIS techniques were used to prepare the thematic maps of the area. Mapping of natural resources is based on GIS software like ERDAS, CAD, etc.

Acknowledgement

The paper is based on the project, it was completed in Institute of Remote Sensing Instruments, Hyderabad during internship.

References

Burrough, P.A. (1987), *Principles of Geographic Information Systems for Land Resource Assessment*, Clarendon Press, Oxford.

Camp, William G. and Thomas B. Daugherty (1995), *Managing Our Natural Resources*, Thomson Delmar, USA.

Clevers, J.G.P.W. (1986), *Application of Remote Sensing to Agriculture Field Truth*, Wageningen Agricultural University, The Netherlands.

Hirani, Kalyani (1991), *Land Use Planning and Geomorphology*, Concept Publishing, New Delhi.

Indian Society of Remote Sensing and Natural Resources Management System (1998), *Remote Sensing for Natural Resources*, Department of Space, Government of India, Hyderabad.

Jensen, J.R. (2004), *Remote Sensing of Environment: An Earth Resource Perspective*, University of South Carolina, Columbia, USA.

Kumar, G. (2005), *Geology of Arunachal Pradesh*, Geological Survey of India, Bangalore.

National Remote Sensing Agency (1995), *Integrated Mission for Sustainable Development: Technical Guidelines*, Department of Space, Government of India, Hyderabad.

—— (1995a), *Report on Area Statistics of Land Use/Land Cover Generated Using Remote Sensing Techniques*, Department of Space, Government of India, Hyderabad.

—— (2002), *Integrated Mission for Sustainable Development*, Department of Space, Government of India, Hyderabad.

Prakash, T.N. et al. (2004), *Earth System Science and Natural Resources Management*, Centre of Earth Science Studies, Thiruvananthapuram.

Verbyla, David L. (1995), *Satellite Remote Sensing for Natural Resources*, Lewis Publishing, Boca Raton, FL.

7

Digitization of Cadastral Maps and their Integration with High Resolution Satellite Data for Land Use/Land Cover Mapping

A Case Study

Reena Devi and L.R. Patel

Man, land and the natural resources supported by land always had an intrinsic and dynamic relationship. Land information is crucial for planning and development. The rapid pace of development can be seen in various cities – that bringing high growth rates also brings with it heavy burdens on land utilization. To ensure that the high growth rates can be sustained, it is imperative to plan for an effective Land Information System, to plan the utilization of its land resources for development, while at the same time protect the environment. Land information refers to any physical, legal, economic or environmental information or characteristics concerning land, water, groundwater, subsurface resources, or air in the state. Land information has been used in a variety of systems over the years; from register of deed tract indexes to surveyors tie sheets or soil surveys. Today, many organizations are using land information through Geographic Information System (GIS).

A Land Information System (LIS) is similar to GIS but is more focused on land records and detailed views of these records. GIS and LIS systems provide tools that support many types of record keeping, analysis and decision-making. Land information is an integral part of government, non-profit, and private sector activities. Adopting

GIS/LIS techniques can advance broader social purposes by making more effective public decisions and by using natural resources in a more optimal way.

Components of LIS

Land information system is an automated development of cadastral mapping. In general there are two main issues of concern for setting up a LIS, viz.:

- Spatial component/Survey data describing the spatial disposition of the parcels in the real world – cadastral maps.
- Non-spatial component describing details such as ownership details, tax value, etc. – legal information.

Land is becoming a scarce resource due to immense agricultural and demographic pressure. Hence, information on land use/land cover and possibilities for their use is essential for the selection, planning and implementation of land use schemes to meet the increasing demands for basic human needs and welfare.

The advancement in the remote sensing technology, particularly in the field of resolution, has brought a revolution in preparing a site-specific developmental action plan. Satellite data gives unbiased information about the objects. There receptivity, multispectral, synoptic coverage, near real time and its relative economy over other methods of study made the techniques better for land use/land cover mapping.

Spatial patterns and variations of different attributes relating to landform, land use and demographic characteristics and their spatial association are easily understandable on the satellite data. Qualitative data retrieved from satellite imageries in the form of maps can be quantified and brought to effective administrative use in planning. Thus, the satellite data are of immense use in the mapping and management of natural resources at cadastral level also.

Definition of Cadastral Map

A cadastral (also spelled cadastre) is a comprehensive register of the real property of a country, and includes details of the ownership, the tenure, the precise location, the dimension (and area) and the value of individual parcel of land. It gives rise to the adjective cadastral, used in public administration, primarily for ownership and taxation purposes. The terminology used for cadastral division may include counties, parishes, riding, hundreds, sections, lots and blocks.

In other words cadastral maps contain methodically arranged information on all properties (parcels or survey number) within a given village or tehsil or district. The scales of cadastral maps are normally one inch = 4 mile or lies in between 1:2500 and 1:8000 (ISRO), in which further details of property namely ownership, land use, its value size, etc., are written for the purpose of collection of revenue from land, based on the security of the property to its own or the legal heirs.

Cadastral maps are an important component of land administration in most countries. In many countries they are taken for granted. Cadastral maps are derived in different ways in different jurisdictions; there are many different perceptions as to what constitutes a cadastral map. Due to historical, cultural and social differences, cadastral maps play very different role in different jurisdictions ranging from being charting or index maps to being the legal determination of parcel boundaries.

Today, cadastral maps have their genesis in systems to manage the land ownership records concerned with the private alienated lands. Since the introduction of the Torren system, subdivisions of private land were charted on index map which had a very low spatial integrity. The integrity of the cadastral system was primarily based on the accurate individual cadastral surveys and plans. Each cadastral survey is usually connected into one or more permanent survey marks in concrete Krebs which is integrated over time into each state coordinate system.

Most computerized cadastral maps in urban areas were initially with the authorities responsible for water, sewerage and drainage in response to their requirement of digital maps to manage all their services. Nowadays, people are concentrating more on site-specific plans that may be either in urban areas or in agriculture, particularly for reclamation purposes, etc. This is possible only when you have a real and reliable database.

Taking this into consideration, a study was undertaken to create cadastral level information with certain objectives in mind.

Objectives

The following objectives have been taken into account in the study:

- Digitization and rectification of cadastral maps;

- Prepare land use/land cover maps through high resolution satellite data;
- Integrate satellite data with cadastral data; and
- Prepare a plan for cadastral information system using GIS.

Study Area

Location and Extent

The study area lies in Ambala district of Punjab which is located between 30°07′ N to 30°34′ N latitude and 76°20′ E to 77°10′ E longitude. The total area of the district is 1,574 sq km.

Two villages in the district were taken as the study area. The locations of these villages are as follows.

The location of Dera is 30°28′50″ N to 30°31′30″ N and 77°10′10″ E to 77°12′50″ E latitude. It is bounded by Jhinwala in the north, Hamidpur in the south, Taprian in the west, and Kala Ambe in the east directions.

The location of the second village, Kurali is 30°23′30″ N to 30°26′ N and 77°3′45″ E to 77°6′ E. It is bounded by Murinpur in the north-east, Alipur in south-east, Salanla in the West and Bukthia village in the south-west directions.

General Description of Study Area

Physiography

Physiographically, the district comprises of a unique topography which is not found in the plains area of Haryana It can be subdivided up to two units, i.e., Siwalik Hills and Piedmont and dissected rolling plains. The villages of the study area fall in the second physiographic unit.

(i) Shiwalik hills: They are located in the northern part of the district with an area 10–15 km in width. The hills are 300–900 m high and they terminate in an undulating sub-mountain tract. These hills are highly eroded.

(ii) Piedmont and dissected rolling plains: This is a transitional unit between the Shiwalik hills and the plains. It is above 30 km in width and is traversed by a large number of rainy season streams of which, Chaggar and the Markanda are the important ones. It is undulating and contains large stretches of silt, sand and pebbles in the beds of numerous seasonal streams, which run down the slopes of the Shiwaliks.

Materials and Methods

The cadastral maps of the study villages were scanned at 200 dpi. The raster data of all the maps thus obtained were vectorized and edited for the missing and non-relevant data. The individual cadastral maps after editing were mosaicked. The boundaries between two adjacent villages were checked and corrected for proper mosaics.

Survey of India topomaps were used as a reference for co-registration among the cadastral, IRS-IC/TD, LISS-IV data. This was achieved by collecting Ground Control Points (GCPs) commonly available on all the datasets by employing geometric rectification technique. The details of the satellite data, other ancillary data, procedures followed are as follows.

Materials Used

(i) IRS 1 C/TD LISS IV Satellite data of 27 November, 2007.
(ii) Survey of India topographical maps.
(iii) The study area falls in toposheets number 53F/1.
(iv) Other ancillary data and related literature/reports
(v) Raster data maps provided by Patwari of the concerned village.

Software Used

AutoCad, ArcEditor, Arc map, and Microsoft word, etc.

Cadastral Data Processing

Digitization

Digitization means acquiring, converting, storing and providing information in a computer format that is standardized, organized and available on demand from a common system. With specialized scanners, manuscripts are converted into compressed digital signals and stored systematically for future reference.

Digitizing is the process of converting features on a paper map into digital format. To digitize a map, a digitizing tablet is used which is connected to the computer to trace over the features. The x, y coordinates of these features are automatically recorded and stored as spatial data.

Digitizing with a digitizing tablet offers another way, besides digitizing "freehand", to create and edit spatial data. It can convert features from almost any paper map into digital features; it can use a digitizer in conjunction with the tools in the editor to create new features or edit existing features on a digital map. The flow chart to show the working of digitization is shown in Figure 7.1.

Figure 7.1

Flow Chart to Show the Working of Digitization

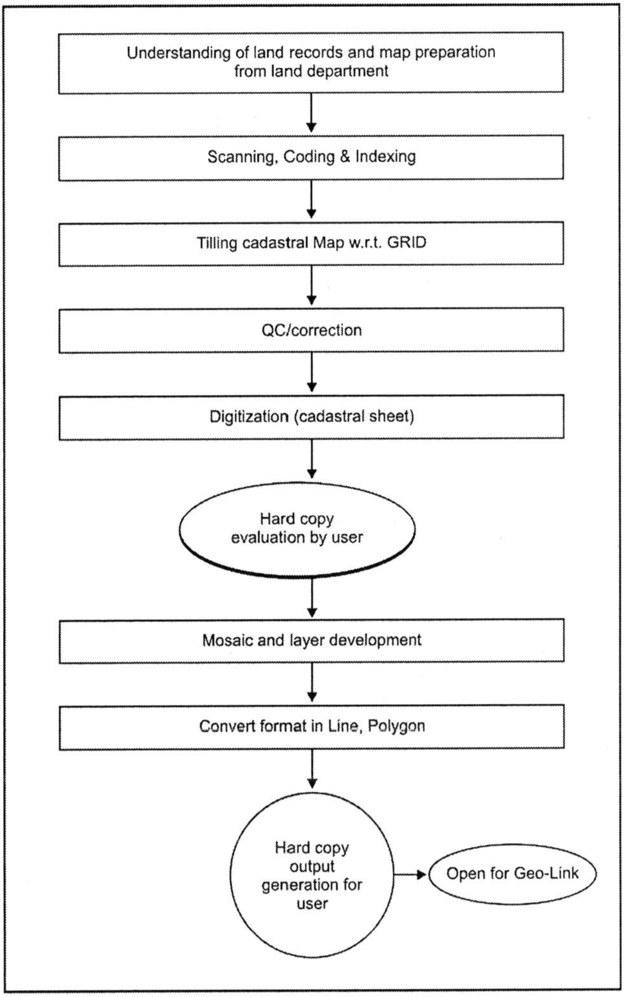

Steps Used During Digitization

(i) The data was loaded as the raw musavi maps of the Dera and Kurali villages provided by Haryana Space Application Centre (HARSAC).

(ii) New shape file for *murrabha* boundary and *murrabha* number were developed. Then one by one each *musavi* was overlaid on the *murrabha* grid.

(iii) Geo-referencing of each *musavi* was done by taking their coordinates. Then each *musavi* was rectified separately.
(iv) Other shape files were developed as *killa* boundary, *killa* number, road line, canal, *burji*.
(v) All the rectified *musavis* were systematically arranged and joined together and were digitized with point, polyline and polygon. Then the margins were edited to view the map accurately.
(vi) Labelling of the map was done by giving titles, legends, scales, and north indicator. Finally, the village map was digitized. The digitized maps of the villages are shown in Plates 7.1 & 7.2.

Geo-referencing of the Satellite Data

Geo-reference box requires two basic inputs:

(a) Uncorrected image.
(b) Master geo-reference image.

Both the images are placed within the same window to facilitate geo-referencing. With proper knowledge, rectification of the imported image is done. Generally the points selected to ensure the basic corrected image has to be visually prominent from the surrounding pixels so that the precision of correction remains higher than 90 per cent.

If correction is performed and run in the first order model then at least three anchor points are required but for second order model minimum 6 to 7 points are required. The images are rectified at a figure from lower to higher numeric proceeding (1, 2 ... n)

When both the images are visualized and they are responding almost on similar location at single click at any of the images then we can consider the uncorrected image.

Steps Used for Land Use/Land Cover Mapping

(i) Base map preparation
(ii) Visual interpretation of data
(iii) Ground truth verification
(iv) Final map preparation
(v) Area calculation

Results and Discussion

The study area covers the part of Ambala district namely, Dera and Kurali villages. These were digitized by following the given methodology. Visual interpretation of land use/land cover classes was done in these villages. Total twelve categories of land use were identified in each village which are: land with scrubs, degraded grazing land, land without scrubs, brick kiln, canal, drainage, river, crop land, fallow land, vegetation, settled area, road line. The village-wise distribution of land use/land cover categories and their maps are given in Table 7.1 & 7.2 and Plates 7.1 & 7.2.

The village-wise description of the land use categories and their areal extent are given below.

Dera

Agriculture Lands

Agriculture is the main occupation in India. In the study area total agriculture land is 1089.51 acre. There are three categories in agriculture lands. Area under various categories of agriculture lands and the distribution of land use in the village is given in Table 7.1.

Table 7.1

Land Use/Land Cover Distribution of Village Dera

S. No.	*Categories*	*Area in acre*
(i)	Agriculture land:	
	Crop land	1047.2
	Fallow land	42.32
	Vegetation	174.24
(ii)	Waste land:	
	Land with scrubs	25.4
	Land without scrubs	33.27
	Degraded grazing land	33.54
	Brick kilns	7.2
(iii)	Water bodies:	
	River	273.8
	Pond	23.05
(iv)	Road	143.9
(v)	Settlement area	74.77

Table 7.2

Land Use/Land Cover Distribution of Village Kurali

S. No.	*Categories*	*Area in acre*
(i)	Agriculture land:	
	Crop land	4389.7
	Fallow land	1213.7

Cont'd...

...Cont'd

(ii) Waste land:	
Land with scrubs	80.71
Land without scrubs	75.79
Degraded grazing land	28.21
Brick kilns	15.13
(iii) Water bodies:	
River	78.02
Pond	4.85
(iv) Road	117.39
(v) Settlement area	40.19

(i) *Crop land:* These are the segments of lands which are very fertile and being used for agriculture purpose. The area under this class is 1047.20 acre. Due to the sufficient irrigation facilities, this category falls under double cropped area.

(ii) *Fallow land:* These fallow land areas are mostly present near villages, agriculture fields or scrubs plantation. The area under fallow land in the Dera village is 42.31 acre.

Waste Lands

These comprise an area of 99.41 acre in the Dera village. The major categories of waste land identified are land with scrubs, land without scrubs and degraded grazing land. The brief description of these categories is as follow:

(i) *Scrubs land:* These areas possess shallow and skeletal soils, at times chemically degraded, extremes of slopes, severely eroded and lands subjected to excessive aridity with scrubs dominating the landscape. These comprise an area of 25.40 acre in Dera village. These are highly eroded unproductive land which can be put to forestation.

(ii) *Degraded grazing land:* These are the lands in non-forest areas that are either under permanent pastures or meadows, which have degraded due to lack of proper soil conservation and drainage measures. These lands are small in size, irregular in shape, scattered and are found close to settlement fringes. It covers an area of 33.54 acre in the Dera village. These are the degraded land around the villages which are not being utilized properly. These lands can be brought under silvipasture practices.

(iii) *Land without scrubs:* This category has a similar description as mentioned in the earlier class excepting that they possess sparse vegetation devoid of scrub and have a thin soil cover. It consists of 33.27 acre of total waste land in the study area. These are uncultivated wasteland which are scattered in patches in the village.

(iv) *Brick kiln:* Mine dumps are those lands where waste debris is accumulated after extraction of minerals. Generally, these lands are confined to the vicinity of mining areas. It includes surface rocks and stone quarries, sand and gravel pits, soil excavation for brick kilns, etc. It covers the area of 7.20 acres in the Dera village.

Rivers, Canals and Water Bodies

Waterlogged land is that low lying land where the water is at/or near the surface and the water stands for most part of the year. The area covered under river is 273.80 acre in the Dera village. The area considered under water bodies is about 23.05 acre.

Roads

It consists of 143.90 acre in the total area of the village. There are two types of roads in the village. First is metalled and the second is unmetalled road.

Settlements

The major part of the settlements lies in the centre of Dera village. There are three settlements in this village and covers an area of 74.77 acre.

Kurali Village

Agriculture Land

In the study area, total agriculture land is 5603.4 acre. There are three categories in agriculture land. Area under various types of agriculture land and the distribution of land use in the village is given in Table 7.2.

(i) *Crop land:* These are the lands which are very fertile and being used for agriculture purpose. The area under this class is 4389.7 acre. The maximum area of the village comes under this category. Due to the sufficient irrigation facilities, this category falls under double cropped area.

(ii) *Fallow land:* These area mostly agriculture lands but are kept fallow due to one or another reason. This category is scattered all over the study area in the agriculture fields. The area under fallow land in the village is 1213.7 acre.

Waste Lands

These comprise an area of 184.71 acre in the Kurali village. The major categories of waste land identified are land with scrubs, land without scrubs and degraded grazing land. The brief description of these categories is as follows:

(i) *Scrub land:* These comprise an area of 80.71 acre in Kurali village. These are highly eroded unproductive lands which can be put to forestation. Plantation of trees will also help in checking the flood in the surrounding plain area.

(ii) *Degraded grazing land:* It covers an area of 28.21 acre in the Kurali village. These are the degraded lands around the villages which are not being utilized properly. These lands can be brought under silvipasture practices.

(iii) *Land without scrubs:* It consists of 75.79 acre of total waste land in the study area. These are uncultivated wasteland which is scattered in patches in the village.

(iv) *Brick kiln:* Mine dumps are those lands where waste debris is accumulated after extraction of minerals. Generally, these lands are confined to the vicinity of mining areas. It includes surface rocks and stone quarries, sand and gravel pits, soil excavation for brick kilns, etc. It covers an area of 15.13 acre in the Kurali village.

Water Bodies

This category includes all ponds, canals/rivers, lanes, etc. The area under ponds consists of 4.85 acre in the Kurali village. The area under river is 82.87 acre.

Roads

It consists of 117.39 acre in the total area of the village. There are two types of roads in the village – metalled and unmetalled. Some main roads are passing between the central parts of village.

Settlements

The major part of the settlements lies in the centre of village Kurali. There are three settlements in the village. Two of them are very small that may be the part of the main settlement area. The total area under this category is 40.19 acre.

Plate 7.1

Cadstral Map of Kurali Village

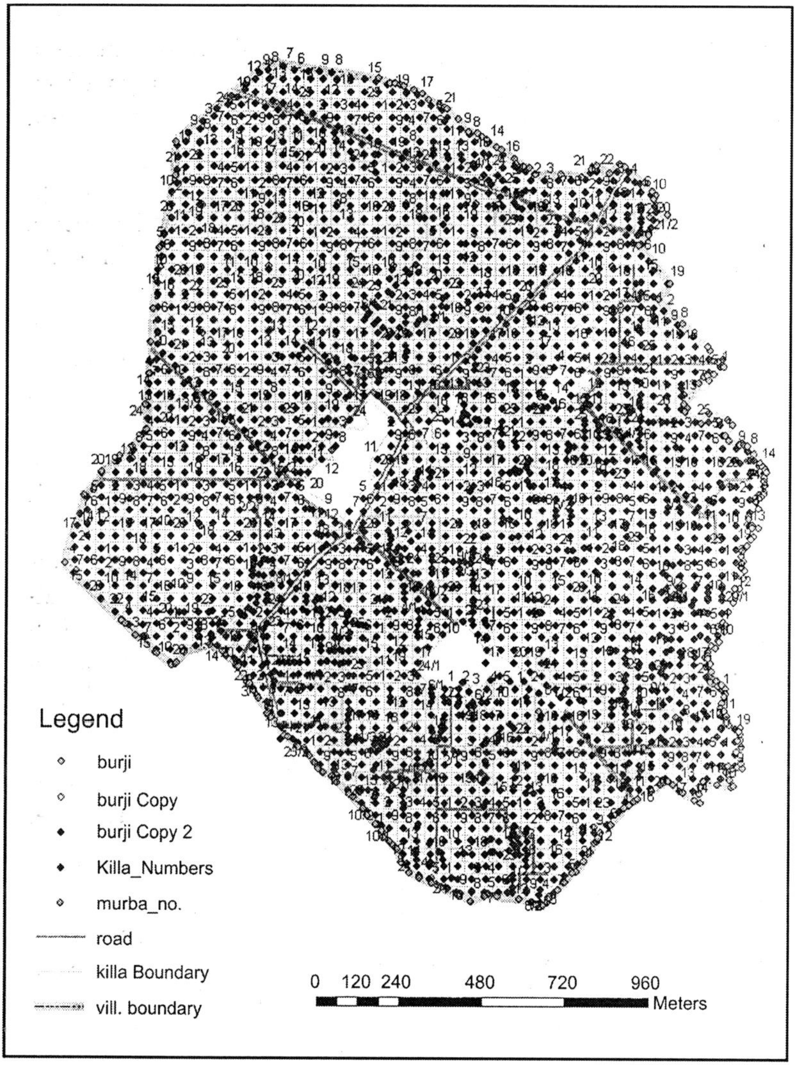

Conclusions

The study shows that the high resolution (LISS – IV) satellite data can be registered with the cadastral maps and a remarkable accuracy has been achieved. Further, it reveals that the cadastral information in the

Plate 7.2

Cadstral Map of Deera Village

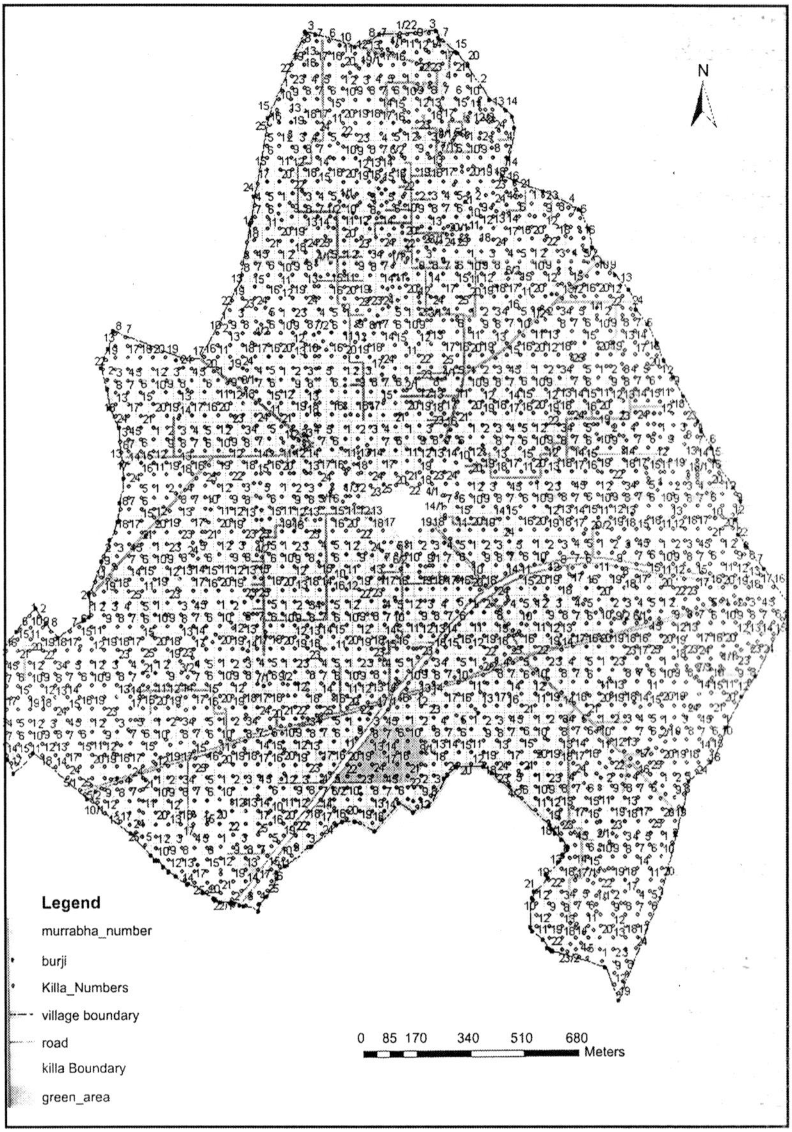

form of maps and records can be updated using IRS-IC/ID high resolution data of appropriate time or cropping season.

It is important that cadastral survey using latest technology may be done for the entire country. Disputed cases in land title should also be decided at the earliest. Computerization process should integrate registration of land titles. Land laws/mutation process should be simplified for easy and fast implementation of computerization.

Various types of land use classes were identified in the study area. These are lands with scrubs, degraded grazing land, land without scrubs, brick kiln, canal, drainage, crop land, fallow land, vegetation, settled area and road line. Based on the work carried out during this study, following conclusions have been derived:

- Scrub lands are found in the study area. This class constitutes 106.11 acre area. The maximum area occupied by this class is found in Kurali village. The land without scrub constitutes 109.06 acre area.
- Total agriculture land of villages is 5603.4 acre. This class included fallow land, vegetation and crop land.
- Degraded grazing land are spread around the villages on the Panchayat land. The grazing land with natural plantations have become degraded due to neglected land management. It covers 61.75 acre of the study area.
- Water bodies cover the total area of 379.72 acre in both the villages.
- Wasteland is an important aspect in land use planning and developmental activities of any area. The total area under waste land is 299.25 acre.

The information generated on cadastral level for these two villages may be very useful for the scientific management of each parcel of land and in developing the Cadastral Information System.

Acknowledgement

The paper is based on project work completed under supervision of Dr V.S. Arya, Sr. Scientist "SG" in the Department of Soil Survey and Land Evaluation in Haryana Space Application Centre, DST, Haryana, Govt. of Haryana.

References

Arya et al. (2004), 'Digital Classification of Current Fallow/Wastelands in Sixteen Districts of Haryana: A Remote Sensing and GIS Approach', HARSAC/TR/01/2004.

Chaurasia, R. et al. (1996), 'Land Use Change Analysis for Agriculture Management: A Case Study of Talwandi Sabo Tehsil in Punjab', *J. Indian Soc. Remote Sensing*, Vol. 31, No. 1, pp. 57–61.

Hooda, R.S. et al. (2006), 'Desertification Status Mapping Using Satellite Data in Dry Sub-humid Region of India: A Case Study', Abstract of the ISPRS TC-IV International Symposium on Geo-Spatial Databases for Sustainable Development, Goa, India, p. 156.

Land Information System Using Geo-Informatics (2008).

Minakshi, R.C. et al. (1999), 'Land Use/Land Cover Mapping and Change Detection Using Satellite Data: A Case Study of Dehlon Block, District Ludhiana, Punjab', *J. Indian Soc. Remote Sensing*, Vol. 27, No. 2, pp. 115–12.

Saha, S.K. and M. Kudrat (1991), 'Selection of Spectral Band Combination for Land Cover/Land Use Classification Using a Brightness Value Overlapping Index (BVOI)', *J. Indian Soc. Remote Sensing*, Vol. 19, No. 3, pp. 141–47.

Singh, Vijay (2000), 'Inventory of Wastelands Mapping of Central Southern Haryana', Statistical Abstract.

UNEP/IRSIC (1991), Global Assessment of Soil Degradation (GLASOD), Nairobi.

Wastelands Atlas of India (2000), Ministry of Rural Development, National Remote Sensing Agency, Department of Space, Government of India.

Wastelands Atlas of Haryana (2006), Haryana Space Application Centre, Department of Science and Technology, Government of Haryana.

8

Palaeo Drainage Mapping by Remote Sensing Techniques

A Case Study of Jhunjhunu District, Rajasthan

P.R. Vyas and Neeraj Chahar

Palaeochannels are typical riverine geomorphic features in a location representing drainage streams, rivers, rivulets which were flowing during the past times and now stands either buried or lost or have shifted due to tectonic, geomorphologic, anthropogenic processes/activities as well as climatic changes. They appear in plain as linear or curvilinear signature and these shapes are identified on remote sensing data products. Synonyms for somewhat similar or related landforms used in lieu of palaeochannels are palaeo drainage, lost river, buried river, buried channel or buried valley.

Palaeochannels or old river courses, buried by subsequent sedimentation, are generally potential reservoirs for groundwater and hence useful for groundwater exploration. In addition to that the palaeochannel identifications are useful in flood mitigation. Soils in this region show more moisture content as compared to the surrounding areas due to the groundwater flow zones. The origin of the paleochannels may be due to the frequent changes in the river. Groundwater prospects in buried channel are excellent because of hydrophilic nature of alluvial and colluvium deposits, retention of foot waters, location being contiguous to the river is mainly responsible for recharge.

Palaeochannels have ideal rock strata for storage. According to prospecting data, the water bearing rock characteristics of these channels are highly porous gravel, coarse-medium sand (in the alluvial fan area), medium fine sand, fine silt (in the river channel zones) and silt (in the delta). Palaeochannels have a relatively convenient source of recharge, as they commonly serve as flood channels or aquaducts and have good paths of infiltration. This is because of the fact that on the surface, they have usually sandy soil or sandy loam. Some may have even sand dunes or river beaches. Most of the palaeochannels are interlinked with present rivers or canals, which is beneficial to lateral recharge of the groundwater.

Palaeochannels can also provide information about the response of catchments and rivers of past climatic changes, with implications for projected changes to climate in the future. Reflect fluvial channels that are filled with high permeability sediments act as preferred pathways for groundwater flow and solute transport. In coastal regions, such palaeochannels can provide a hydraulic connection between freshwater aquifers and the sea, facilitating saltwater intrusion landward or freshwater discharge offshore.

Water Resource Potentiality of Palaeochannels

Palaeochannels considered as lost drainage are still having under flow of groundwater evidenced by abnormally high pumpage discharge, fresh to low saline quality in comparison to water occurring in adjoining non-riverine environment and are also comparatively older in character. This is a premise on the basis of which efforts are being made to trace the course of the Saraswati river – one of the very famous long rivers lost in the Thar Desert. Such palaeochannels are being looked upon as possible reservoir of groundwater. If such basins are further identified and integrated, it is possible to rejuvenate it through recharge by rainwater harvesting as well as by canal system. Such measures may initiate a blue revolution in the Thar Desert. It is obvious that palaeochannels have comparatively better prospects of water storage in comparison to adjoining non-fluvial zone. They have a great potentiality of holding greater volume of water and as such on the basis of similar analogy palaeochannels may also be looked upon for water to meet the demand during the drought situation. However, there is need to identify, delineate and estimate the sedimentary thickness of all such palaeochannels for precise planning of water resources development and utilization programme to effectively counter the menace of drought.

Quality of Water in Palaeochannel

Further, in order to assess the quality of the groundwater in the palaeochannels a total of 37 water samples were collected from dug wells confined to different lithological units. The geochemical analyses of water samples are carried out and TDS, pH, CO_3, HCO_3, Cl, SO_4, NO_3, F^+ were calculated. Total hardness as $CaCO_3$ and Ca^{++}, Mg++, Na+, K+, SiO_2 and B were determined. The epm per cent values and atomic ratio values of Na+/K+ Cl+/HCO_3 +, 01-F, Cl-/SO_2; Na+/Mg+; Mg+/Ca+ and total alkalinity of these samples were calculated. The percentage values of all these samples were plotted on the piper diagram which reveals that most of the water confined in abandoned playas and rocky pediplain areas is NaCl type. The sand dune field areas in $NaHCO_3$, type of undergroundwater. The quality of water is potable and fit for irrigation purposes as well. No doubt it holds much hope for future as huge untapped reservoir of fresh water resources.

It has been observed by the geologists that the presence of a large number of segments of palaeochannels in the region using IRS I-C data. All these palaeochannels are presumed to be the relics of a river system that drained western Rajasthan at different times during the Quaternary, like the Ghaggar in north-western Rajasthan. Haryana and also the Sindh Province in Pakistan, attest the presence of erstwhile fluvial regimes in the Thar Desert region.

Signature of Palaeochannel

It has been observed that the palaeochannels exhibiting linear, curvilinear, and loop-like forms and appearing with typical black ribbon like stripes and loops in black and white products and characteristic reddish stripes and loops in colour coded formats can be seen in that area. These signatures appear due to moisture enrichment and vegetation along old courses and buried river systems.

Palaeochannels present in the study area were extracted from final geomorphologic map. These were delineated from satellite image by considering combination of factors like linear/curvilinear/meandering courses, channel like shape, high moisture, linear depressions, vegetated regions and association with fluvial landforms. These channels show a discontinuous drainage pattern, trending in NE – SW and NNE – SSW directions; their width ranges between 100–300 m channels close to present-day river course and have comparatively high moisture as compared to others.

These palaeochannels were occupied by reworked sand heaped in the form of dunal ridges. Presently, signatures of these palaeochannels have been obliterated from the ground due to levelling/sand removal for cultivation. Their sinuous shape, high moisture and concentrated vegetation cover helped in the identification from the multi-data satellite data.

Satellite data has advantages of synoptic view of multi-spectral and multi-temporal nature. Useful landforms like palaeochannels in arid regions are sometimes as wide as 6–8 km and their length may be of the order of hundreds of kilometres. It becomes intricate to visualize their presence due to agricultural fields, or soil cover, e.g., palaeochannel of lost Sarasvati observed in the northern parts of Rajasthan and Haryana along the bed of current Ghaggar river.

Palaeochannels – Prospects (Utility)

In view of the adverse prediction of the various international and national agencies about the availability of the water resource from 2025 to 2050 and the surfacing water scarcity conditions in Jhunjhunu district, the delineation and identification of palaeochannels may be highly useful from the following point of view:

(i) Source of the groundwater, in general of good quality.
(ii) Possible locales for rainwater harvesting.
(iii) Possible locales for artificial recharge using excess of canal water, etc.

These palaeochannels are not only the abode of potable water, but also posses high volume of storability which can be utilized by recharging them by catching each and every droplet of the rainfall-runoff component. Also, imported water through existing canals, may provide a relief to drought-affected people and livestock. Such measures may be adopted for drought proofing in this state These palaeochannels are not merely the imprints of the lost channels but are the river's movement which need to be understood in greater details and possibly stands out to provide not only needed water resources to the society but a lesson too to modern humanity for taking proper care of their existing riverine environment and to take suitable measures to prevent them from turning into palaeochannels.

Basics of Remote Sensing

Remote sensing has transformed the understanding the processes of nature. The analysis, application and interpretation of remotely sensed images have brought significant advances in a wide range of fields from climatology, geology, agriculture and oceanography to urban planning, environmental monitoring, etc. The advent of satellites has allowed the acquisition of information about the earth and its environment. Sensors on the earth resources and meteorological satellites provide information on surface vegetation cover, its seasonal variation, surface morphological structures, sea surface temperature, wave heights, patterns and dynamics of clouds, etc.

Study Area

Jhunjhunu district covers an area of 5.928 sq km which is 1.73 per cent of the total area of Rajasthan. The district is located in the north-eastern part of the state. It is surrounded by Churu on the north-western side, Hissar and Mahendragarh of Haryana in the north-eastern part and by Sikar in the west, south and south-eastern part. Jhunjhunu district comprises of the total population of 19,13,099. The literacy rate of the district is 73.61 per cent. About 68.24 per cent of the population is engaged in agriculture. The main crops of the district are mustard, wheat, bajra, gram and barley. The district has 3 industrial areas with 3,742 small scale and 4 large and medium scale units.

The Jhunjhunu district shines unique among all the districts of Rajasthan. Most part of the district is semi-desert. The Aravali ranges are embracing the south-eastern part of the district. The huge and magnanimous copper fields are in the bowl of these ranges in Singhana and Khetri suburbs.

Location

The district is situated in the north-eastern part of the Rajasthan State in between Latitude 28°13′ N, longitude 75°4′ E and its altitude is 358 metres. It is surrounded by Churu on the north-western side, Hissar and Mahendragarh of Haryana in the north-eastern part and by Sikar in the west, south and south-eastern part.

The district is famous for its lush green valleys and beautiful natural scenes captivate the tourists. Lohargal, the holy shrines of pilgrimage, lies in the lap of these ranges. Some of the most popular tourist destinations in the district include Alsisar, Bagar, Bissau,

Chirawa, Churi Ajeet Garh, Dundlod, Gangiasar, Jhunjhunu, Kajara, Khetri, Kirori Ji, Lohargal, Mahansar, Malsisar, Mandawa, Mukandgarh, Narhar, Nawalgarh, Parasrampura, Bara, Pilani, Singhana, Surajgarh and Udaipurwati.

Physical Aspect

The district is spread over 5,928 sq km area. Most of the district consists of rolling hills and the remaining part nearer to south-eastern border contain some offshoots of Aravali range which enters the district in the extreme south of Udaipurwati tehsil and extends up to Singhana and Khetri in the east. Western part of the district has vast stretches of sand dunes and sand sheet. Shifting sand and active sand dunes are the main hazards to cultivation. The drainage in the district is inland and is related to Kantli river system, which is a non-perennial river. This river enters into Jhunjhunu district in Udaipurwati tehsil and after passing through the district, ultimately enters Churu district and becomes extinct there itself. Besides this river, there are four major streams, namely Sohan, Chandrawati, Udapur Loharagh Ki Nadi, and Sukh Nadi. There is no lake in the district but sacred tanks are in existence for bathing purposes.

Data and Methodology

The application of remote sensing data for palaeochannel delineation is based on broad principles namely, spatial and genetic relationships between surface material and underlying palaeochannel fills.

The study is mainly based on the satellite imageries and ground truthing with the help of the global positioning system (GPS). The satellite imageries and their details used in this work are presented in Table 8.1 and the salient features of the IRS-1C (LISS III) sensors are shown in Table 8.2.

Data Used

Table 8.1

Details of the Satellite Imageries

Satellite	*Sensor*	*Path*	*Row*	*Date*
IRS-1C	LISS - III	94	51	22-Mar-06
IRS-1C	LISS - III	94	52	9-May-06
IRS-1C	LISS - III	95	51	14-May-06
IRS-1C	LISS - III	95	52	7-Jun-06

Table 8.2

The Characteristics of IRS-1C (LISS-III) Sensors

Salient Characteristics	*IRS-1C (LISS-III)*
Spectral channels	Channel 1, 520–590 nm Channel 2, 620–680 nm Channel 3, 770–860 nm Channel 4, 1155–1700 nm
Special resolution	23.5 m (Channel 1–3) 70.5 m (Channel 4)
Swath	148 km
Quantization	8 bit
Repeat cycle	24 days
Launch date	December 28, 1995

Toposheet Used

Survey of India Toposheet 1:50.000
Jhunjhunu – 44/P, 53/D, 45/M

Softwares Used

(i) *ERDAS Imagine*: ERDAS Imagine is a standard image processing software. It has a high degree of administration flexibility, new space-saving features, many import and export utilities, greater mobility for fieldwork, superior mosaicking functions and advanced 3D visualization and scene creation capabilities. It also includes numerous image enhancement techniques for the purpose of better image interpretation and extracting more reliable information.

Key Features Summary

(a) Superior mosaicking functions.
(b) Advanced 3D scene visualization.
(c) Better integration between GIS layers and images. Virtual GIS for scene building.
(d) Extensive Vector data editing tools.
(e) GPS receiver live link
(f) Hyper linking capability in the ERDAS Imagine Data Viewer.
(g) Various techniques for spatial spectral and radiometric enhancements.

(ii) *ARC/INFO Workstation 8.2*: ESRI's ARC/INFO Version 8.2 marks another step for professional GIS. Arc/Info Workstation 8.2 enables the user to improve spatial database construction, advanced visualization and analysis, advanced cartography, and spatial database management.

Image Analysis and Palaeochannels Identification

Palaeochannels Identification in Jhunjhunu District

Palaeochannels has been traced on satellite image using digital image processing techniques using ERDAS (8.5) Imagine software. Enhancement techniques were applied for contrast enhancement, histogram equalization; principal component analysis band rationing, edge enhancement, filtering, etc., have been performed on the IRS-1C LISS III data to identify the palaeochannels in the Jhunjhunu district of Rajasthan. The extent of channel continuity has been traced on band 1 and band 2 of IRS-1C LISS III imagery by the typical light gray to white tone for the dry part and dark gray tone for the part with moisture and vegetation. On false colour composite, the dry sections are indicated by their white to light greenish tone and the moist and vegetated sections by pinkish red tone.

If we look at the drainage pattern of Jhunjhunu district we will find south-eastern part of the district is having a good drainage system. It actually originates from Aravalis and Kantli at present. The drainage in the district is inland and is related to Kantli river system, which is a non-perennial river. This river enters into Jhunjhunu district in Udaipurwati tehsil. After passing through the district, it ultimately enters Churu district and become extinct there itself. Besides this, there are four major streams, namely Sohan, Chandrawati, Udapur Loharagh Ki Nadi, and Sukh Nadi and few sacred tanks are in existence for drainage purposes. Palaeochannels of Kantli river believed to be a tributary of Saraswati river have been identified. The causes of the Kantli river shifts have been worked out. The signature of Palaeochannels on satellite data has been observed to be misleading when correlated with field data. Use of aerial photos has improved the identification of palaeochannels. Kantli river has been identified using digital enhancement techniques.

Palaeochannels has been traced on satellite image using digital image processing techniques using ERDAS (8.5) Imagine software. Various enhancement techniques such as contrast enhancements, histogram equalization principal component analysis, band rationing, edge enhancement, filtering, etc., have been performed on the IRS-1C LISS III data to identify the palaeochannels in the Jhunjhunu district of Rajasthan. The extent of channel continuity has been traced on band 1 and band 2 of IRS-1C LISS III imagery by the typical light grey to white tone for the dry part and dark grey tone for the part with moisture and vegetation. On false colour composite the dry sections are indicated by their white to light greenish tone and the moist and vegetated sections by pinkish red tone. The studies identify palaeochannels along the tributaries of the Jawai-Sukri and the Luni rivers in the Thar Desert.

The aerial extension of palaeochannel was delineated on 1:50,000 scale through digital image processing of remote sensing (IRS 1C LISS III) data. The false composite colour (FCC) of the area was generated after applying principal component analysis. The FCC so obtained, was then visually interpreted for precisely delineating the course of palaeochannel. Analysis of the FCC reveals existence of a linear feature starting from Udaipurwati tehsil. Kantli river is spread over near villages Chirawa, Udaipurwati, Buhana, Alsisar, Bagar, Bissau, Khetri and Mandawa Alsisar, Dundlod, Gangiasar, Jhunjhunu, Kajara, Kirori Ji, Lohargal, Mahansar, Malsisar, Mukandgarh, Narhar, Nawalgarh, Parasrapura, Bara, Pilani, Singhana, and Surajgarh. With a view to narrowing down the target width of buried channel, existing bore holes, hydrographs and meteorological data were also studied in conjunction with the surface data. This resulted in further narrowing down the effective target area along the course of the channel and it was further reduced after carrying out detailed geophysical survey. The remote sensing data interpretation of the buried channel was correlated with geomorphologic geohydrological and other related evidences. It has been observed that in general, the channels occupy the dissected pediments and get disorganized or extinct in the aeolian plain or dome complex. The interpretation from the satellite data is based on the light grey to white tone due to sand cover and the alignment of vegetation cover. The vegetation cover appears as red tone on the false colour composite, whereas a distinctive vector layer has been prepared identifying the road, drainage and palaeochannels of the Jhunjhunu district.

Summary and Conclusion

Remote sensing is being widely used to reveal the palaeochannels in the desert areas through various digital image-processing techniques applied to interpret the satellite images. The present disorganized state of many streams in the Jhunjhunu district area of Rajasthan has been attributed by many investigators to change in climate during

Quaternary period, and in particular the Holocene period which are attributed to the present disorganized state of many streams in the Thar Desert of western Rajasthan. The analysis of satellite images of the area, however, indicates a number of lineaments across the plains. In most of the cases, drainage disappearance are related to these lineaments. The pattern of drainage disorganization suggests that climate could at best be an associated factor, but not the primary one for the disappearance of the palaeo drainage system. The pattern of drainage disorganization suggests that the climate could at best be an associated factor but not the primary one.

The ground truth result shows a sign of availability of water as the presence of good crop was seen in the field. After verification from farmers and other sources it was found that part of Kantli river is spread over near Chirawa. There are overwhelming evidences from satellite imagery interpretations, as well as from archaeological, geological and geophysical evidences that the rivers with well-integrated drainage once flowed as palaeochannel through several parts of Rajasthan including the Jhunjhunu district.

These palaeochannels are not only the abode of potable water but also posses high volume of storability which can be utilized by recharging them by catching each and every droplets of the rainfall-runoff component. The imported water, through existing canal, may provide a relief to drought affected people and livestock. Such measures need to be adopted for drought proofing in the study area. These palaeochannels are not merely the imprints of the lost channels but are the river movements which need to be understood in greater details and possibly stands out to provide not only needed water resource to the society but also the lesson to modern humanity for taking proper care of their existing riverine environment and to take suitable measures to prevent them from turning into palaeochannels. Remote sensing along with surface and subsurface geophysical methods are highly appropriate tools to delineate the geometry of the channels and to assess the groundwater storage potentiality of the palaeochannels.

Palaeochannels are being convincingly looked upon as a possible reservoir of groundwater with vast storage capacity. There is a need to identify and to establish the storage potentialities of these landforms, particularly in such regions where groundwater/water emergency situations are anticipated. The artificial recharge of

Plate 8.1

Location Image Showing Palaeochannels, Drainage, Roads, Jhunjhunu District

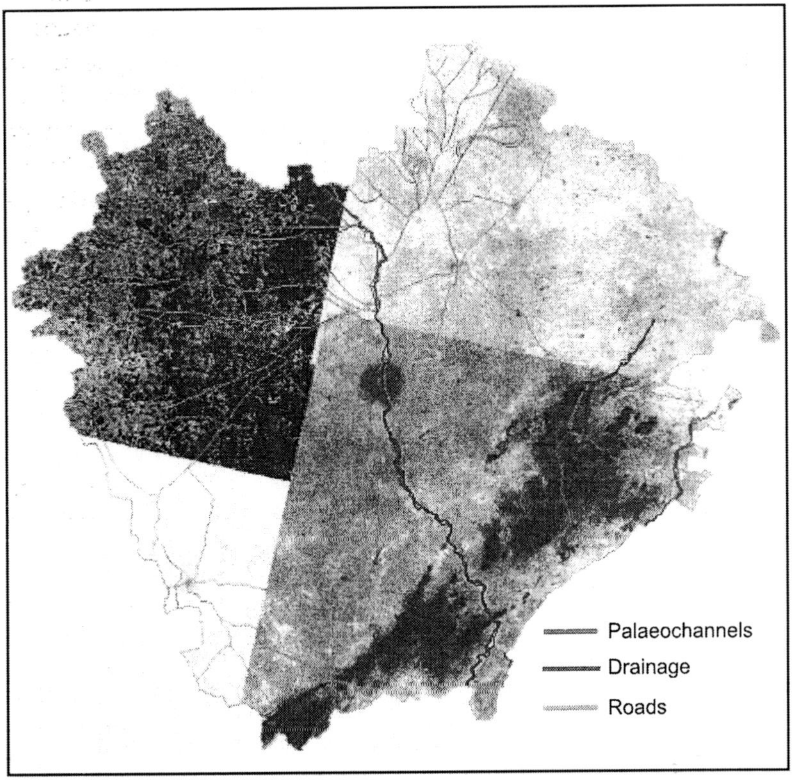

palaeochannels to its full capacity may provide adequate long-term storage of groundwater, which may be used to mitigate the impact of drought. In fact, flash flood occurred due to storm water, lean/non-supply of water through canal/transponder aquifer/river and seismic effect, thus reducing the unsustainability of freshwater supply during emergency situations to the minimum.

Palaeochannel studies suggest that all of the sand and gravel layers of the alluvial fans and the sand layers of the palaeochannels be used as shallow-well extraction zone. Large amounts of shallow freshwater should be extracted to evacuate the storage capacity so that an underground reservoir may be constructed through artificial recharge. This will not only alleviate the problems of drought, flooding and salinity, but also simultaneously make thorough use of our water resources.

But there are some critical technical problems in using underground reservoirs for storage. These are:

(i) How to speed up the rate of infiltration, especially that of flood waters (or major amounts of transferred water).

(ii) How to enlarge the storage capacity: that is how to gradually enlarge the freshwater storage of the palaeochannels while reducing the saltwater storage between channels.

Acknowledgement

The paper is based on project work, completed by Ms Neeraj Chahar under supervision of Prof. P.R. Vyas (Internal Guide) and Dr M.P. Punia, Head, Remote Sensing Division, B.M. Birla Science and Technology Centre, Jaipur (Raj.)

References

Bakliwal, P.C. and S.B. Sharma (1980), 'On the Migration of the River Yamuna', *Journal of Geological Society of India*, Vol. 21, September, pp. 461–63.

Bakliwal, P.C. and A.K. Grover (1988), 'Signature and Migration of Sarasvati River in Thar Desert, Western India', *Rec. Geol. Surv. Ind.*, Vol. 116, Nos 3–8, pp. 77–78.

Bakliwal, P.C. and S.M. Ramasamy (1987), 'Lineament Fabric of Rajasthan and Gujarat, India', *Records of the Geological Survey of India, Jaipur*, Vol. 113, No. 7, pp. 54–64.

Ghose, B., A. Kar and Z. Hussain (1978), 'Comparative Role of the Aravali and the Himalayan River Systems in the Fluvial Sedimentation of the Rajasthan Desert', Paper presented at the Symposium on Tertiary and Quaternary Climate and Environmental Changes, Tenth International Congress of Anthropological and Ethnological Sciences, December.

Kar, A. (1994), 'Lineament Control of Channel Behaviour During the 1990, Rajasthan', *Journal of the Geological Society of India*, Vol. 32, pp. 522–26.

—— (1998), 'Possible Neotectonic Activities in the Luni-Jawai Plains, Rajasthan', *Journal of the Geological Society of India*, Vol. 32, pp. 522–26.

Karanth, K.R. and K. Seshu Babu (1978), 'Identification of Major Lineaments on Satellite Imagery and on Aerial Photographs for Delineation of Possible Potential Groundwater Zones in Penukonda and

Dharmapuram Taluka of Ananthapur District', Proc. of Joint Indo-US Workshop on Remote Sensing of Water Resources, NRSA, Hyderabad, pp. 188–97.

Palanivel, S., A. Ganesh and T. Vasantha Kumaran (1996), 'Geohydrological Evaluation of Upper Agniar and Vellar Basins, Tamil Nadu: An Integrated Approach Using Remote Sensing, Geophysical Well Inventory Data', *J. Indian Soc. Remote Sensing*, Vol. 21, No. 4, pp. 217–27.

Prithvi Raju, N. (1980), *Geomorphic Studies in Sarada River Basin, Visakhapatnam District, Andhra Pradesh, India*, Awarded Ph.D. Thesis, Andhra University, Visakhapatnam, India.

Raju, K.C.B., G.V.K. Rao and B.J. Kumar (1985), 'Analytical Aspects of Remote Sensing Techniques for Groundwater Prospecting in Hard Rocks', Proc. of the Sixth Asian Conference on Remote Sensing, pp. 127–32.

Ramasamy, S.M. (1999), 'Neotectonic Controls on the Migration of Sarasvati River of the Great Indian Desert', in *Vedic Sarasvati, Evolutionary History of a Lost River of Northwestern India*, Memoir Geological Society of India, Bangalore, No. 42, pp. 153–62.

Sankar, K., M.S. Jegatheesan and A. Balasubramanian (1993), 'Hydrogeological Studies in the Kanyakumari District, Tamil Nadu', Regional Seminar on Groundwater Development Problems in Southern Kerala, Proc. Vol., pp. 12–18.

Snelgrove, A.K. (1979), 'Migrations of the Indus River: Pakistan in Response to Plate Tectonic Motions', *Journal of the Geological Survey of India*, Vol. 20, August, pp. 352–403.

Tiwari, O.N. (1992), 'Fallibility of Palaeochannels as Groundwater Potential Zones in a Part of Thar Desert', *Journal of the Geological Society of India*, Vol. 40, July, pp. 70–75.

Yashpal, Sahai Baldev, R.K. Sood and D.P. Agarwal (1980), 'Remote Sensing of the Lost Sarasvati River', *Proc. Ind. Acad. Sci.* (Earth and Planet Sci.), Vol. 89, No. 3, pp. 317–31.

9

Groundwater Prospecting in Suk Tawa Sub-Watershed Using Remote Sensing and GIS Techniques

Veena Charan, Amarjeet and Chanchal Choudhary

Groundwater being a valuable resource in today's world it needs proper evaluation and management for overall development within a region for its proper use. Water is one of the most important renewable resources hidden within the earth's crust. Increase in population and growth of industries at a rapid rate have put pressure for the development of the regions and also for the survival of the population. To overcome this major problem it is essential for proper development, management and conservation of groundwater resources.

Groundwater occurrence at any place on earth is not a matter of chance but a consequence of the interaction of climatic, geologic, and physiography factors. In India, the groundwater occurs in three types of geological formations:

- *Unconsolidated:* This covers about 30 per cent of land area; primarily the Indo-Gangetic plain. The lithology comprises sand, gravel, pebbles, etc., which can store large quantities of groundwater;
- *Semi-consolidated:* This covers only about 5 per cent of the geographic area. The lithology includes primarily semi-consolidated sandstone, formations of Mesozoic and Tertiary age; and

- *Consolidated:* This constitutes 65 per cent of the geographical area comprising hard-rock formation. These include crystalline rocks – granites, gneisses, etc.

In the hard rock areas, the movement and occurrences of groundwater depends mainly on the secondary porosity and permeability resulting from folding, faulting, fracturing, etc. Porosity, the volume of available pore spaces in rocks, determines the amount of water which can be held in storage. Permeability determines the ease with which water moves through the pores and fractures. The most obvious structural features that are important from the groundwater point of view are lineaments. The lineaments are linear or curvilinear features and play a vital role particularly in geomorphic and structural analysis. The lineaments like joints, fractures, etc., developing generally due to tectonic stress and strain, provide important clue on surface features and are responsible for infiltration of surface run-off into sub-surface and also for movement and storage of groundwater. Photolineaments generally represent the surface traces of fractures in bedrocks, projected more or less vertically upwards to the erosion surface by various mechanisms.

Remotely sensed data from satellite provide quick useful baseline information on the factors controlling the occurrence and movement of groundwater, like geology (lithology/structure), geomorphology, land use/land cover, etc. Remote sensing techniques have proved superior than other conventional groundwater surveys because remote sensing has the advantage of repetitive coverage of the area, is time saving, economical and most successful than the conventional surveys.

Objectives

The main aim of the study is to evaluate the groundwater prospects in Suk Tawa sub-watershed using remote sensing and GIS tools. To achieve this aim, work is carried out with the following objectives:

- To generate the lithological map of the study area.
- To generate the geological structure with lineaments of the area.
- To map geomorphological features.
- To prepare the maps for integration of the lithology, geomorphology and structure in ARC/INFO GIS environment to generate final groundwater prospecting maps.

Study Area

The study area selected for the present project work is Suk Tawa sub-watershed in Hoshangabad district, Madhya Pradesh State. The watershed lies between 22°15′N to 22°28′N latitude and 77°40′E to 77°56′ E longitude.

Physical Background of Study Area

Hoshangabad district is situated in the southern part of Madhya Pradesh and covers an area of about 6,705 sq km. It falls in Survey of India degree sheet No. 55 B&C, 55F, 55J and 55G and lies between latitudes 21°54′N to 23°00′N and longitudes 76°46′E to 78°43′E.

Hoshangabad is bounded by Dewas district in the north-west, Sehore and Raisen districts in the north, Narsinghpur district in the east, Chindwara district in the south-east. Betul district is in the south and east and Nimar district in the west. The districts lie south of Narmada River and have a rectangular shape. The present study area, i.e., Suk Tawa sub-watershed is located in the southern part of Hoshangabad district.

The climate of Hoshangabad district is normal. Average height from the sea level is 331 metres. The average maximum and minimum temperatures are 32° and 19°C respectively. Overall, the district has a temperate climate. Average rainfall in this area is about 134 cms.

Methodology

Data Used

Remote sensing techniques have been mainly used for groundwater prospects evaluation. Satellite data of IRS-P6 LISS-III, FCC (Plate 9.1) and other collateral data like Survey of India (SoI) toposheets (55F/15 and 55F/11), existing maps and reports were used for studying the physical features and preparing the base maps. LISS-III data have been interpreted to map the major faults, fractures/lineament, geomorphology, lithology, etc., of a selected area for identifying prospective groundwater zones.

Satellite	*Sensor*	*Date of Acquisition*
IRS – P6	LISS – III	23 Dec 2003
IRS - P6	LISS - III	16 Apr 2005

Plate 9.1

Ground Water Prospect Map for Suk Tawa Sub-Watershed

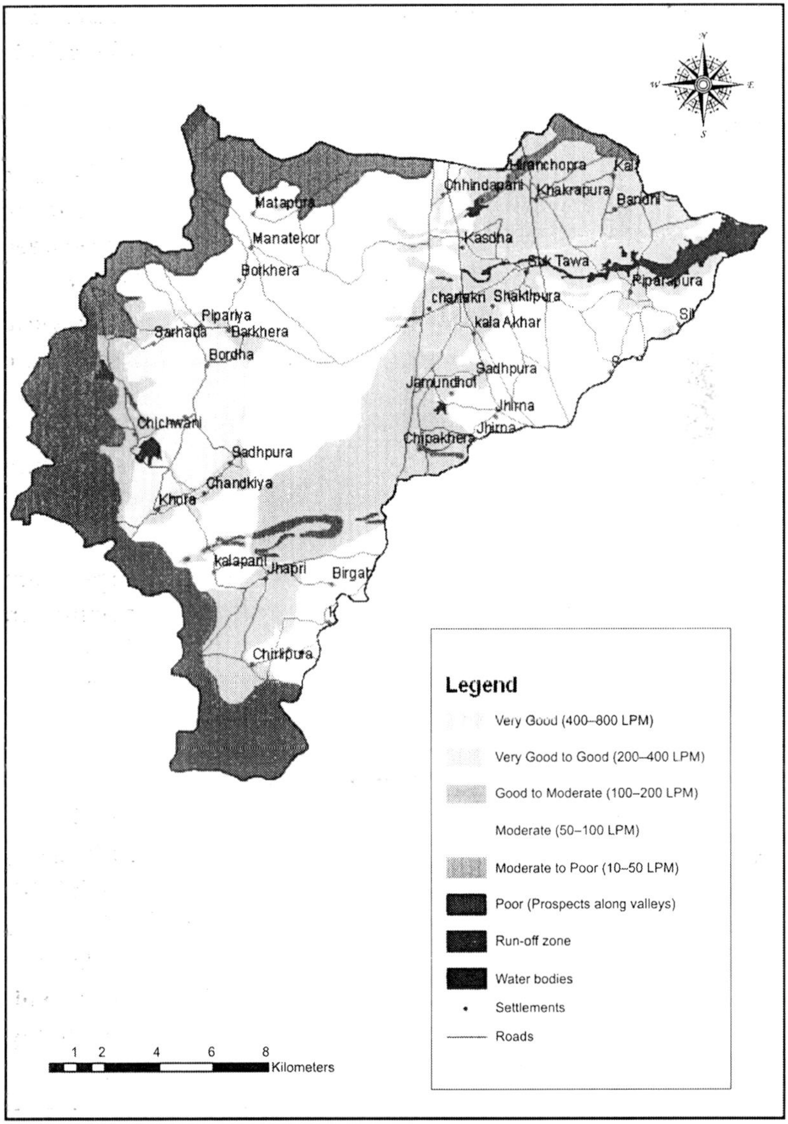

The Groundwater Prospects Mapping

The board methodology used for preparing the groundwater prospects map on 1:50,000 scales comprised the following four main steps:

(i) Preparation of base map, lithological map, geomorphological map, lithological map and structural map overlays based on the visual interpretation of satellite image.
(ii) Field visit for checking the interpretation and collecting the additional hydrogeological information.
(iii) Incorporation of field observations in the lithological, structural, geomorphologic, and hydrological map overlays.
(iv) Preparation of groundwater prospects map by integrating the lithological, structural, geomorphologic and hydrological map overlays and transferring the details on to the base map.

Preparation of Base Map

The base map has been prepared by extracting features from the existing maps and updated by using latest satellite imagery. The features such as water bodies (both perennial and ephemeral), canal, national highway, and other metalled and unmetalled roads, railway and settlements have been included in the base map.

Preparation of Lithological Map Overlay

The synoptic view and multi-spectral nature of the satellite imagery help in discrimination and mapping of different lithologic units. Geological mapping has been carried out mainly based on visual interpretation of satellite image adopting deductive approach by studying image characteristics and terrain information in conjunction with a prior knowledge of general geological setting of the area. The tone (colour) and land form characteristics combined with relative erodibility, drainage, soil type, land use/land cover and other contextual information observable on the satellite image have been used for differentiating different geological formations.

Preparation of Structural Map Overlay

The synoptic coverage provided by the satellite imagery enable mapping regional structures which is difficult in conventional ground surveys due to scanty rock exposures, soil cover and lack of continuous observations. Lineaments representing the faults, fractures, shear zones, etc., are the most obvious structural features interpretable on the satellite imagery. They control the occurrence and movement of groundwater and their significance in groundwater exploration has been proved beyond doubt.

They appear as linear to curvilinear feature on the satellite imagery and are often marked by the presence of moisture, alignment of vegetation, straight stream/river courses, alignment of tanks/ponds, etc. These characteristics often indicate presence of lineaments and have been used for mapping lineaments in the study area.

Preparation of Geomorphic Map Overlay

The synoptic view of satellite imagery facilitates better appreciation of geomorphology and helps in mapping of different landform, and their assemblage. The photo-interpretation elements, such as tone, texture, shape, size, location, association and other criteria like physiography, genesis of the landforms, nature of rocks/sediments associated geological structures, etc., have been used for identification of different landforms/geomorphic units. The existing maps provide topographic information required to interpret the geomorphology from the satellite imagery. Initially, the entire image has been classified into major zones, i.e., hills and plateaus, piedmont zones and plains considering the physiography and relief as the criteria. Then, within each zone, different geomorphic units have been mapped based on the landform characteristics, their areal extent, depth of weathering, thickness of deposition, etc., the geomorphic units mapped from satellite imagery have been checked in field during ground truth exercises.

Preparation of Hydrological Map Overlay

Satellite imagery provide excellent information on hydrologic aspects like stream/river courses, canals, major reservoir, lakes, tanks, etc. Based on visual interpretation of satellite data all the above information can be derived and mapped.

Ground Truth and Data Collection

Before proceeding to the field for ground checks, proper planning has to be made regarding the type of data to be collected and the locations. For this purpose, based on the road network, the doubtful area marked with dashed lines while interpreting lithology, landforms, faults, fractures, etc., which need to be verified in the field have been identified and noted in the field book. Similarly, the locations of the problematic area have been marked on the map overlays for checking them on the ground for further details.

During the field visit, the doubtful areas in the different map overlays have been verified/checked on the ground and necessary corrections/modifications have been incorporated in the maps.

Wherever the confirmed boundaries have been drawn, they have been randomly checked to ascertain the correctness of interpretation.

Integration of Thematic Layers and Field Data

The data from lithological, structural, geomorphological and hydrological map overlays have been integrated and transferred on to the base map to prepare the groundwater prospects map on 1:50,000 scale. The following steps are involved in the integration:

(i) Generation of coverages of the above map overlays in the ARC/INFO environment. The lithological and geomorphological map overlay has polygon, line and point features, structural map overlay has and point features and base map overlay has polygon, line and point features.

(ii) Editing the coverages to remove errors, such as dangle nodes, overshoots, undershoots, etc.

(iii) Building the topology after the errors are brought within the tolerance limits.

(iv) Labelling the features as per the standard codes/symbols for representing lithological/geomorphological/structural/hydrological/ base map units.

(v) Union of geology and geomorphology layers.

(vi) Overlaying the structure, hydrology and integrated lithology and geomorphology layer on the base map. This represents the groundwater prospects map.

(vii) Evaluating each hydrogeomorphic unit in term of: (a) the groundwater prospects, (b) the type of wells suitable, (c) the yield range, (d) problems/limitations, etc.

Result and Discussion

In the present project work, groundwater prospects mapping using remote sensing technique and GIS tool has been undertaken for Suk Tawa sub-watershed. The groundwater prospecting maps prepared by GIS based integration of lithological, structural, geomorphic (landforms) and hydrological themes mapped from satellite images provided all the information required for proper evaluation of groundwater conditions. Such maps form a very good data base and help the user departments in identifying suitable zones/sites for planning recharge structures to improve the sustainability of drinking water sources.

Geology of the Study Area

Geologically, the area consists of litho-units of Mesozoic to Quaternary eras. A study of the geology of this region and an understanding of the geological history are important for exploring the groundwater resources.

The lithology map of the study area has been prepared using the visual interpretation of satellite images, taking the existing GSI maps as reference. The sub-watershed of Suk Tawa lies in southern Satpura hill range.

The litho-stratigraphic units of the study area are described in the following paragraphs:

(i) *Quaternary:* Quaternary alluvial deposit occupy major part of the area. Apart from this, a number of dykes trending in different directions cut across the Gondwana litho-units.

  (a) Alluvium: Alluvium is made up of gravel and fine silty deposits with sand and layers of pebbles derived from the neighbouring rocks.

  (b) Dykes: The dykes range in the length from few hundred metres to few kilometres with width ranging from few metres to few hundred metres. Most of dykes trend in east to west and north-east to south-west directions.

(ii) *Deccan Trap (Satpura Group):* Deccan traps occur as lava flows overlying granite gneisses of Achaean to Proterozoic age. The trap is melanocratic, dense and medium grained. The basaltic lava flows of Deccan trap are well exposed in the southern and south-western part of the area. These flows grouped under Satpura group are mainly of Aa type and non-porphyritic to porphyritic to mega-porphyritic in nature. The thickness of individual flows varies from 15m to 47m. The Satpura group comprises of 18 to 21 basaltic flows.

(iii) *Gondwana Supergroup:* The Gondwana sequence belonging to the Gondwana basin of central India comprises of Bijori, Pachmarhi and Jabalpur formations. The Bijori formation is exposed as a broad band. It comprises olive and buff coloured clays and shales. alternating the massive sandstone. The Pachmarhi formation consists of thick beds of coarse to granular, white arenite or quartz-wake, separated by lenses or thin layers of clast supported conglomerate and thin red clay bands. The youngest Gondwana sequence is represented by Jabalpur formation. It consists mainly of massive sandstone alternating with white clays.

Criteria for Lithologic Discrimination

The important characteristics of common rock types manifested on the remote sensed data are briefly discussed below.

Sedimentary Rocks

The most conspicuous characteristic of sedimentary rocks is its stratifications, which gives banded appearance on the remotely sensed images/photographs. Commonly occurring sedimentary rocks are sandstones, shales and limestones. Horizontally bedded sandstones being resistant to weathering, form massive hills and plateaus with steep side slopes. Dipping sandstones form cuesta hills, homoclinal and hogback ridges.

Shales, being incompetent rocks, form subdued topography. In humid climate, they form gently rounded hills with very gentle slopes. The drainage in shales is external and is mainly dendritic. They have fine drainage texture and badland topography develops in arid regions. Cultivation is generally practised over shales because of the presence of moisture in soils and their low-lying nature. Shales show darker tone compared to sandstones because of more surface moisture.

Igneous Rocks

Igneous rocks are divided into intrusive and extrusive igneous rocks. Plutonic igneous rocks are characterized by their massive (unbedded) and homogenous nature, and their discordant relationship with the country rocks. Granitic rocks are more common in nature compared to other plutonic igneous rocks, and are characterized by the presence of hummocky topography, i.e., they form dome-like hills surrounded by vast undulating plains.

Dolerites, which generally occur as dykes form linear to curvilinear ridges and have discordant relationship with the country rocks. However, when the surrounding rocks are more resistant, dykes form depressions. Sills are difficult to identify on the images as they are intruded parallel to the bedding of sedimentary rocks and can be misinterpreted as one of the sedimentary layer.

Extrusive igneous rocks can be easily identified based on the presence of volcanic land forms, such as lava cones, lava flow having tongue like pattern, etc. Thick deposit of basalt, form plateaus and mesa/butte type of topography.

Metamorphic Rocks

Metamorphic rocks are comparatively more difficult to identify on the image/photograph – commonly occurring metamorphic rocks are gneisses, schists, phyllites, slates and quartzite. These rocks except quartzite, have subdued undulating topography and manifest regional foliation/schistosity/cleavage in the form of parallel lineation called trend line, on the images. The banding in metamorphic rocks are short, irregular and numerous compared to long, continuous and evenly spaced banding in sedimentary rocks.

Geomorphic Units of Study Area

Geomorphology is the study of landforms and landscapes including the description, classification, origin, development and history of planetary surfaces. The lithologic unit (rock type) and geomorphic unit (land forms) are the fundamental units of geology and geomorphology respectively.

The geomorphological processes are generally complex and reflect interrelationship among the variables such as climate, geology, soil and vegetation.

The delineated geomorphic units in the study area have been broadly grouped into denudational and fluvial origin based on their genesis (Amarjeet, 2006).

Denudational Landforms

The fluvio-denudational geomorphological processes are actively involved in landscape reduction process. The physio-chemical and biological weathering formed the different Denudational landforms in the study area. The denudational landforms in the study area include highly/moderately/lowly dissected plateaus, pediments, pediment inselberg complex, pediplains and weathered pediplains (Amarjeet, 2006).

Highly/moderately dissected plateaus mainly form the run-off zones, whereas the lowly dissected plateaus and pediments have very poor groundwater prospects. Weathered pediplains have well to moderate groundwater prospects.

Fluvial Landforms

The fluvial depositional landforms are formed under the influence of corrosive and erosive dynamics process due to high drainage density and deep weathering and are responsible for the development of various depositional landforms. The fluvial landforms in the study area are alluvial fans. Alluvial fans are good for groundwater occurrence and therefore form very good groundwater prospective zones.

Structure of the Study Area

Linear features of geological interest, called lineaments, representing joints, fractures, faults, etc., are significant for groundwater prospecting. The lineaments on the remote sensing data can be identified mainly based on the linear anomalies, such as tonal anomalies, straight drainage courses, alignment of water bodies in a straight line, linear vegetation anomalies, etc. Lineaments of the study area have been interpretated using the above-mentioned anomalies.

A number of minor and major lineaments are present in the Suk Tawa sub-watershed. The lineaments are trending in north-west to south-east, east to west and north-east to south-west directions (Plate 9.1). In general, the zones along the lineaments and their intersections have good groundwater prospects.

Groundwater Prospects of the Study Area

The integration of lithology and geomorphology has given the different hydro-geomorphological units with varied groundwater prospects (Plate 9.1). The different hydro-geomorphological units and their groundwater prospects are described below:

(i) Alluvial fans over alluvium: These fluvial landforms occur at the western part of the watershed at the foot of the highly/moderately dissected plateaus. These are formed due to the deposition of thick alluvium brought down the plateaus by streams. They have very good groundwater prospects (400–800 LPM).

(ii) Highly/moderately dissected plateau over sand stone, shale, clay, siltstone and conglomerate: These denudational landforms occur in western part of the watershed. They have very poor to nil groundwater prospects (run-off zone).

(iii) Lowly plateau over clay, shale, siltstone and they too have very poor groundwater prospects: These denudational landforms occur in the eastern part of watershed. They are poor in groundwater prospects.

(iv) Pediments over sandstone, pebble bed, shale, clay, silt stone and conglomerates: These landforms occur in the central part of watershed. They are moderate in groundwater prospects (50–100 LPM).

(v) Pediplains over clay, shale, gritty sandstone, siltstone, pebbles bed, dykes and sills: These fluvial landforms occur in the centre part of the watershed area. They are moderate in groundwater prospects (50 – 100 LPM).

Table 9.1

Groundwater Prospects for Suk Tawa Sub-Watershed and their Areal Statistics

Groundwater Prospect Category	*Area in Sq Km*
Very Good (400–800 LPM*)	16.90
Very Good to Good (200–400 LPM)	76.80
Good to Moderate (100–200 LPM)	26.80
Moderate (LPM)	46.70
Moderate (50–100 LPM)	91.30
Moderate to Poor (10–50 LPM)	56.10
Poor (Prospects along Valleys only)	23.10
Poor to Nil (Run-off Zone)	35.7

* *LPM: Refers to liters per minute, a valumetric flow rate.*

Conclusion

The study of groundwater prospects of Suk Tawa sub-watershed has been carried out using remote sensing and GIS tools. The following are the conclusions of study:

The Suk Tawa sub-watershed geologically is represented by rock formations of Mesozoic to Quaternary age. Major part of the study area is occupied by the sandstone, shale, silt stone, gritty stone litho-units of Gondwana super group. Small areas in the southern and western margin are covered by Deccan basalts. Quaternary alluvium deposits are found along foothill zones and along streams.

The broad geomorphic unit's landforms of Suk Tawa sub-watershed include the highly/moderately dissected plateau, lowly dissected plateau, pediments, pediplains with varied thickness of weathered mantle and alluvial fans.

The geology and geomorphology together with other factors like climate, control the groundwater prospects. The study has shown that the remote sensing techniques have tremendous potential for mapping of geology, mainly the structures and geomorphological features of the area. GIS is found to be a powerful tool for integrating these various information and generation of groundwater prospects map.

Suk Tawa sub-watershed has very good to good groundwater prospects in alluvial plains. The groundwater prospects are moderate in pediplains over shales, whereas the prospects are moderate to good in the weathered pediplains over sandstone litho-units. The groundwater prospects are very poor to poor in pediments and pediment's inselberg complex area. The moderately dissected/highly dissected plateau form mainly the run-off zones.

Acknowledgement

The paper is based on project work, completed by Ms Veena Charan under supervision of Prof. P.R. Vyas (Internal Guide) and Dr G. Sreenivasan, Scientist, RRSSC, DOS-ISRO, Nagpur.

References

Amarjeet, K. (2006), *Geomorphological Mapping in Suk Tawa Sub-Watershed*, Project Report submitted to JRN Rajasthan Vidyapeeth University, Udaipur.

Belward, Alan S. and C.R. Valenzuela (1991), *Remote Sensing and Graphical Information System for Resources Management in Developing Countries*, Kluwer Academic Publishers, London/Boston.

Bhagvan, S.V.B.K. and V. Raghu (2005), 'Utility of Check Dams in Dilution of Fluoride Concentration in Ground Water and the Resultant Analysis', *Environmental Geochemistry and Health*, Vol. 27, pp. 97–108.

Drury, S.A. (1987), *Image Interpretation in Geology*, Allen and Unwin, London.

Freeze, R. Allen and John A. Cherry (1979), *Groundwater*, Prentice-Hall, Englewood Cliffs, New Jersey.

Jensen, J.R. (2004), *Remote Sensing of the Environment: An Earth Resources Perspective*, University of Carolina, Columbia, USA, pp. 117–233.

Joseph, G. (2004), *Fundamentals of Remote Sensing*, Cambridge University Press, Cambridge, UK.

Kumar, V.K. and T.R. Martha (2004), *NNRMS Bulletin*, Department of Space, Government of India, Bangalore.

Lillesand, T.M., R.W. Keifer and J.W. Chipman (2004), *Remote Sensing and Image Interpretation*, John Wiley & Sons, New York.

Matthess, G. (1982), *The Properties of Groundwater*, John Wiley & Sons, New York.

National Rural Drinking Water Mission Methodology Manual for Preparation of Groundwater Prospective Zone Maps (1999), NRSA, Department of Space, Hyderabad.

NRSA (1995), *Integrated Mission for Sustainable Development*, National Remote Sensing Agency, Department of Space, Government of India, Balanagar, Hyderabad.

—— (2000), *Rajiv Gandhi National Drinking Water Mission: Technical Guidelines for Preparation of Groundwater Prospects Maps*, National Remote Sensing Agency, Hyderabad.

Reddy, M.A. (2001), *Remote Sensing and Geographical Information System*, B.S. Publications, Hyderabad.

Reddy, P.R., K.V. Kumar and K. Seshadri (1996), 'Use of RS-1C Data in Groundwater Studies', *Current Science*, Vol. 70, No. 7, pp. 600–05.

—— (1991), *Use of IRS-1C Data in Groundwater Studies*, National Remote Sensing Agency, Balanagar, Hyderabad.

RRSSC (2005), *Remote Sensing Studies for Diamond Exploration in Panna RP Area, Madhya Pradesh*, Regional Remote Sensing Service Center, Nagpur.

Sabins, S.F. (1997), *Remote Sensing Principles and Interpretation*, 3rd Edition, W.H. Freeman & Co., New York.

Sahai, B., A. Bhattacharya and V. Shegde (1991), *IRS-1A Application for Groundwater Targeting*, Remote Sensing for National Development.

Sreenivasan, G., A.K. Gupta and J.R. Sharma (2000), 'Selection of Sites for Water Harvesting and Groundwater Recharge Structure Using Remote Sensing and GIS Approach', Compendium of Papers presented in National Seminar organised by Social Policy Research Institute, Jaipur.

Vyas, P.R. (1996), *Application of Remote Sensing for Natural Resource Management: Ecological Crisis and Environmental Protection*, Agrotech Publications, Udaipur.

10

Prospecting Groundwater Resources by Remote Sensing and GIS Techniques

Study of Sagwara Block, Dungarpur District

Somendra K. Sharma and Pankaj Rawal

Water is an important natural resource on the earth and plays a vital role in every aspect of life. The sources of consumable water are mainly groundwater, rain and surface water bodies like river, ponds and lakes, etc., but the competitors for their consumption are too many like domestic sector, agricultural sector, infrastructure sector and industrial sector, etc. The agricultural sector is the major consumer of water and land resources. Ever-increasing pressure of population coupled with poor management of land and water resources are putting tremendous stress on our fragile environment. The limited availability of natural resource has made it mandatory to use them optimally, so that the need of each sector can be fulfilled with justification. The scenario of fast growing world compelled the human civilization to employ efforts that maintains the developments with sustained utilization of natural resources. With increasing demand of water by the growing world population, it has become necessary for mankind to look for additional sources, as well as to conserve the present available resources.

Objectives

The fundamental objective of the study is to generate groundwater prospecting maps, which help in planning and various resources development programmes. The broad objectives are the following:

(i) To prepare thematic maps, i.e., Base map, Drainage map, Geomorphology map, Lithology map, Structural map, Hydrology map and Groundwater prospects map.

(ii) To collect, collate and analyse both primary and secondary data at block level and prepare related maps.

(iii) To integrate thematic and secondary data to get groundwater prospects map.

Study Area

Dungarpur district has been selected as the study area for the project work. It is situated in southernmost part of Rajasthan between 23°20′N to 24°01′N latitudes and 73°21′E to 74°23′E of longitudes. Its borders are attached with Banswara and Udaipur districts in the east and north respectively, while it adjoins the State of Gujarat in the south and west. Dungarpur district occupies 1.13 per cent of the total area of Rajasthan. Most part of the district is hilly. The overall land productivity is rated to be low for the whole district with somewhat better conditions found in its southern and western corners.

The district, though fairly open in the south and east, is interspersed with stony hills covered with low jungle of cactus, jujube trees and salar (*Bosswellina serreta*) together with several other varieties of shrubs and trees which require neither a deep soil nor moisture. The eastern part slopes down towards the basin of the Mahi river and consists of plain and level cultivated area. In hilly tract, the cultivated area is confined to the valley and ground between the hills where the soil is rich with alluvium and can be irrigated from wells and tanks. Sagwara town is centrally located in the block and is well connected with other major towns of the regions.

Sagwara block is located in the south of Aspur block between 23°32′N to 23°48′N latitudes and 73°46′E to 74°10′E longitudes along with Mahi river forming the eastern boundary of the block, having a total geographical area of 611.72 sq km (including the town).

Lithology

The lithological assemblage of the Dungarpur district forms southern continuity of Ajmer-Udaipur metamorphic belt which belongs to the oldest known rocks of the Mangalwar complex (Pre-Aravalis) and Aravali Super Group of rocks, with intrusive ultramafic suite of rocks.

Ultramafics

The Rikhabdeo ultramafic suite of rocks can be traced in this district as a southern continuity in the Dungarpur and Simalwara blocks. In between, it is cutt off by a north-north-west trending major fault in the southern part of Dungarpur block. This zone is widest (around 2 km) in the south-west of Manpur village and is characterized by its distinct darker tones. These rocks also occur as isolated patches in the north-western, northern and western parts of the area. This ultra-mafic suite of rocks consists of serpentine talc-carbonate rocks and chlorite schists, characterized by green colour. They are generally alphinitic but sometimes exhibit discernible crystals of titanite, garnet, chromite, magnesite, amphiboles, etc. They usually show well-developed schistisity (nearly north-south) which is parallel to the rocks they are introduced in.

Quaternary

The Quaternary deposits consists of valley-fill and alluvium. The alluvium consists mainly of riverine sediments (sand, silt and gravel) which are mainly confined to narrow zones along the banks of Som river and larger streams. In rest of the area, most of the stream courses are with valley-fill material of varying depths, consisting of coarse elastics.

Tools And Techniques

Data Used

Slope map, hydro-geomorphological map, drainage, groundwater potential are the thematic maps prepared for the study. Features on a toposheets or other analog documents were used through scanned maps. They were imported and, geo-referenced then digitized by 'on screen digitization' in ERDAS Imagine. Toposheets of the area of study were used like 46 I/1, I/2, E/13, E/14, ETM FCC and published literatures were also used.

The ground truth verification is also done as and when required at various stages. The local knowledge of the area has been taken into consideration at the time of interpretation. All drainages are taken out from toposheets and further modified by satellite data. The groundwater prospect of the delineated geomorphic units are evaluated using available hydrologic characteristics and aquifer parameters.

Satellite Data

For better results Landsat ETM image is used. Landsat current flagship sensor, the enhanced thematic mapper (ETM), was placed in orbit in April 1999, with the successful launch of Landsat 7. This system designed by NASA is operated by NOAA and USGS. The ETM was also used for a 15 m panchromatic channel. Swath width remains at 185 km, and the system is characterized by improvements in accuracy of calibration, data transmission and other characteristics.

Implementation of Methodology

Assigned Rank and Weight to Different Thematic Layers

Mostly, five thematic layers were used in this plan; these are drainage, geomorphology, slope, geology and structure. For further processing assigned Rank and Weight to thematic layers were made, according to characteristics of the study area. Rank and weight of all layers are shown in Tables 10.1 to 10.5.

Table 10.1

Assigned Rank and Weight to Drainage Order (D.O.)

Drainage Order	*Rank*	*Weight*	*Rank × D.O. × Weight*
1	1	1	1
2	1	1	2
3	1	1	3
4	1	6	24
5	1	7	35
6	1	8	48
7	1	9	63

Table 10.2

Assigned Rank and Weight to Lithology

Lithology Unit	*Rank*	*Weight*	*Rank × Weight*
Quartzine	2	1	2
Ultramafic Rocks	2	4	8
Gneiss	2	6	12
Phyllite	2	9	18

Table 10.3

Assigned Rank and Weight to Geomorphology

Geomorphic Unit	*Rank*	*Weight*	*Rank × Weight*
Structural Hills	4	1	4
Denudation Hills	4	3	12
Burried Pediplain	4	6	24
Floor Plain	4	7	28
Dry Water Body	4	7	28
Pediplain	4	7	28
River	4	8	32
Perenial River	4	8	32
Water Body	4	8	32
Valley-fills	4	9	36

Table 10.4

Assigned Rank and Weight to Slope Category

Slope Category	*Rank*	*Weight*	*Rank × Weight*
Very steep slope	3	1	3
Moderately steep to slope	3	4	12
Strongly sloping	3	5	15
Moderately sloping	3	6	18
Gently sloping	3	7	21
Very gently sloping	3	8	24
Nearly level	3	9	27

Table 10.5

Assigned Rank and Weight to Structure

Structure Type	*Rank*	*Weight*	*Rank × Weight*
Lineament	5	1	5
Structural Trends	5	9	45

Conversion into Raster (Pixels) Data

For completion of the calculation, thematic layers have been converted into raster (pixel) format for further processing through feature to raster convert function in Arc Map.

Preparation of Integrated Maps by Using Formula

Integration of various thematic maps was performed by overlaying one theme map over the other, through cross operation of raster maps in ArcEditor software. The cross operation performs an overlay of two raster maps by comparing pixels at the same position in both maps and keeping track of all the combinations that occurred between the values or classes in both maps. The geo-reference of input raster maps was kept the same. During the cross operation, combinations of class names, identifiers or values of pixels in both maps were listed and the number of pixels occurring in this combination was computed. Rank and weight of all the thematic layers were multiplied by each other and divided by total of rank values. The results were stored in the output cross map and in a cross table. The output cross table obtains an identifier domain, with the same name as the output map. This domain contains items, which were combinations of the class names, IDs, groups, name or values of the first input map.

Figure 10.1

Schematic Diagram Showing Methodologies

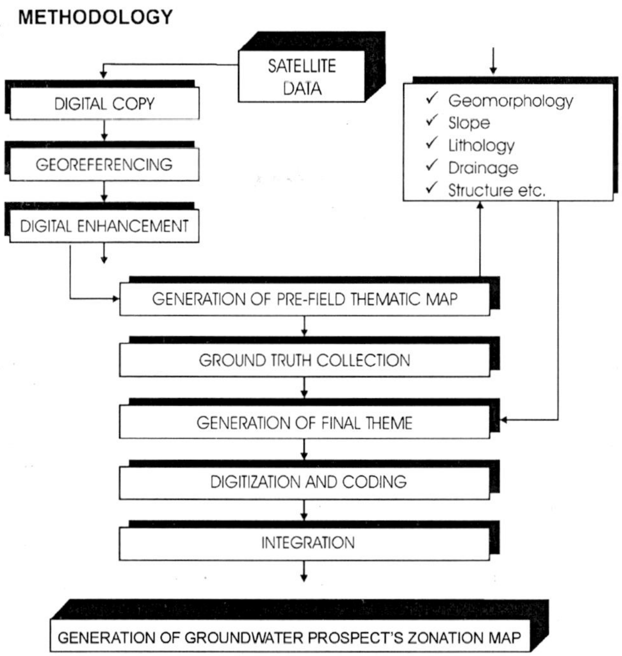

Groundwater Resource Planning

Preparation of groundwater resource planning involves study of different thematic maps and quick visit to the study area for verification. A number of spot observations were made by covering almost all types of landforms, slope, groundwater potential and quality, rainfall and climate, present land use, etc. The land characteristics as mapped in the respective theme maps were recorded along with the present land use. The cross map, of all themes shows all combinations, which were possible through integration procedure.

The main hydrogeological units occurring in Sagwara are granite gneisses of banded gneissic complex phyllites and schists and ultrabasic rocks of Aravali Super group. Sagwara block mostly has phyllites and schists aquifers (349 sq km) followed by granite gneisses (56 sq km). Fluctuation ranges from 0.51 to 5.83 m in phyllites and schist zone and 1.85 to 2.84 m in granite gneisses aquifers with an average fluctuation of 4.07 m. In Sagwara, undulating topography plays a dominant role in potential areas of groundwater recharge. Most of the rainwater flows down along slopes as surface run-off towards the drains, ephemeral streams and rivers and a small amount of it is stored in the topographical depressions such basins, valley-fills and flood plains. Because of the limited extent of such geomorphic units, the overall groundwater prospect is limited. The major hydrogeological zone is phyllite/schist with poor groundwater prospect zone and the depth at which the water is available, ranges between 4–10 m. Overlaying the gneissic rocks of the Pre-Aravalis are the quartzite, forming high rising ridges in the area which have narrow crests. The quartzite is massive, silicified, well-jointed and depicts bedding traces having steep dips. The groundwater condition in valley-fill is good to very good; in irrigated and flood plain, moderate to good; in burried pediplain area, poor to moderate; in structural and denudational hills, poor to very poor. Most of the lineaments are confined to the valley regions and these lineaments are acting as pathways for groundwater movements.

Conclusion

Remote sensing and GIS technology proved to be very useful for groundwater prospecting and water resource planning activities where the potential gap is still to be filled up. It is concluded that the study conducted makes efficient use of groundwater resources and its conservation for future use. It helps in better land use planning and regional development.

The geomorphological interpretation of the ETM FCC imagery are described on the basis of hydro-geomorphological interpretation coupled with Survey of India toposheets and existing geological

Plate 10.1

Groundwater Potential Zone Map of Sagwara Block

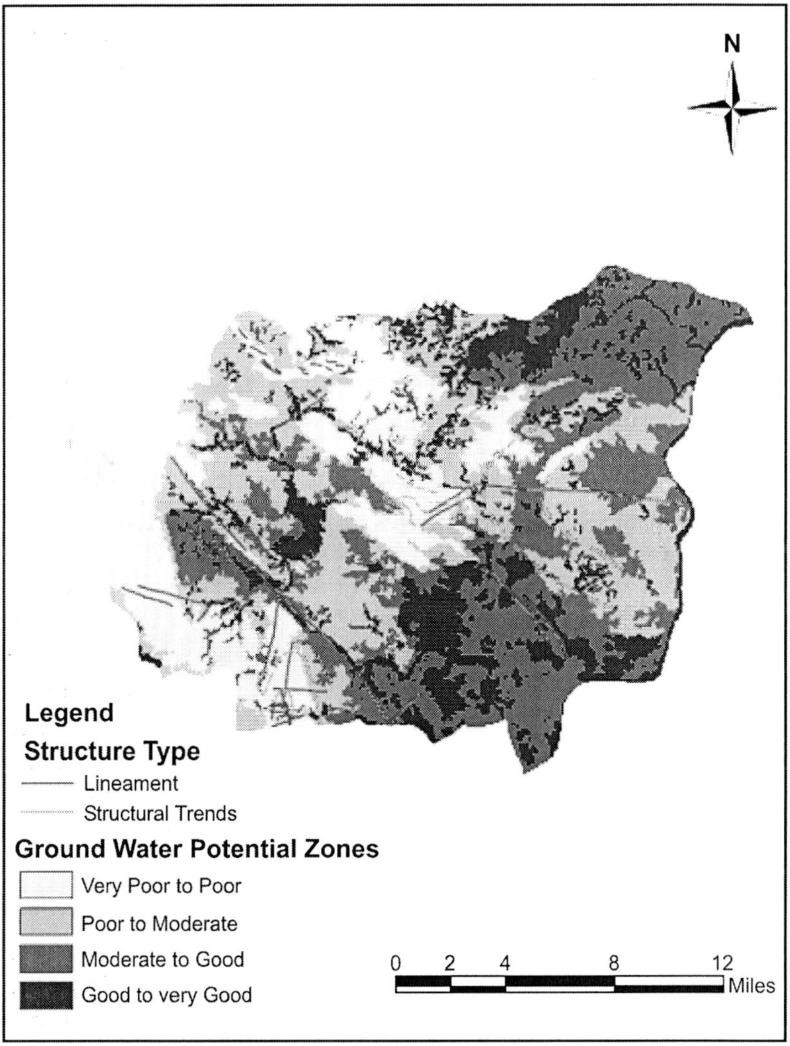

information. The study also helps in the identification of potential groundwater zones.

Groundwater Prospect

The groundwater prospect in the block is moderate to good. The terrain is essentially formed of phyllites/schistose rocks, gneisses and ultramafics, which do not form very good aquifer in general except in few places where secondary porosity (faults, fractures intersections of joints, foliation and cleavages) permits percolation of rainwater. There is also very thin soil cover on this lithology except flood plain and valley-fills areas where depth of Quaternary cover is only sufficient to warrant shallow to medium depth wells and low duty tube wells.

The quality of groundwater is generally suitable for irrigation and drinking purposes except in a few localities where fluoride is high. The yield of dug wells range from 30,000–45,000 lpd in phyllites/schistose and 30,000–40,000 lpd in gneisses and ultramafic.

Acknowledgement

The paper is based on project work, completed by Mr Somendra K. Sharma under supervision of Prof. P.R. Vyas (Internal Guide) and Dr K.R. Khichar, Sr. Scientific Officer, B.M. Birla Science and Technology Centre, Jaipur.

References

Belward, Alan S. and C.R. Valenzuela (1991), *Remote Sensing and Graphical Information System for Resources Management in Developing Countries*, Kluwer Academic Publishers, London/Boston.

Bhagvan, S.V.B.K. and V. Raghu (2005), 'Utility of Check Dams in Dilution of Fluoride Concentration in Ground Water and the Resultant Analysis', *Environmental Geochemistry and Health*, Vol. 27, pp. 97–108.

Dubey, Alok, P.K. Singh and Alok Choudhary (1999), *Morpho-Geological Analysis for Groundwater Resource Evaluation and Management*, Project Report submitted to IIRS, Dehradun.

Jensen, J.R. (2004), *Remote Sensing of the Environment: An Earth Resource Perspective*, University of Carolina, Columbia, USA, pp. 117–233.

Joseph, G. (2004), *Fundamentals of Remote Sensing*, Cambridge University Press, Cambridge, UK.

Kumar, V.K. and T.R. Martha (2004), *NNRMS Bulletin*, Department of Space, Government of India, Bangalore.

Lillesand, T.M., R.W. Keifer and J.W. Chipman (2004), *Remote Sensing and Image Interpretation*, John Wiley & Sons, New York.

Matthess, G. (1982), *The Properties of Groundwater*, John Wiley & Sons, New York.

National Rural Drinking Water Mission Methodology Manual for Preparation of Groundwater Prospective Zone Maps (1999), NRSA, Department of Space, Hyderabad.

NRSA (1995), *Integrated Mission for Sustainable Development*, National Remote Sensing Agency, Department of Space, Government of India, Balanagar, Hyderabad.

—— (2000), *Rajiv Gandhi National Drinking Water Mission: Technical Guidelines for Preparation of Groundwater Prospects Maps*, National Remote Sensing Agency, Hyderabad.

Padmavathy, A.S., N. Yagaranjan, P. Thangavel and M.G. Chandrashekhar (1993), 'Check Dam Site Selections Using GIS Approach', *Adv. Space Res.*, Vol. 13, No. 11, pp. 123–27.

Rao, D.P. (1995), *Remote Sensing for Earth Resources*, Association of Exploration Geophysicists, Osmania University, Hyderabad, pp. 153–69.

Reddy, P.R., K.V. Kumar and K. Seshadri (1991), *Use of IRS-IC Data in Groundwater Studies*, National Remote Sensing Agency, Balanagar, Hyderabad.

RRSSC (2005), *Remote Sensing Studies for Diamond Exploration in Panna RP Area, Madhya Pradesh*, Regional Remote Sensing Service Centre, Nagpur.

Sabins, S.F. (1997), *Remote Sensing Principles and Interpretation*, 3rd Edition, W.H. Freeman & Co., New York.

Part – 3

Application of Remote Sensing and GIS Techniques for Environmental and Regional Planning

11

Application of Remote Sensing and GIS in Land Use and Land Cover Analysis

P.R. Vyas and R.P. Sharma

The study of land use cover is very useful in present context for the purpose of planning of any area. Importance of land as economic resource, whether rural or urban, is long recognized. Efforts are therefore directed to achieve the best possible use of land under a given situation. Several tools and techniques have been developed from time to time for efficient utilization of land. Mapping organization of database and assigning its use or zoning are major aspects of land management. Studying temporal changes in the use of land, assessing the future needs and promoting the land use accordingly are the core activities of land-use planning. The natural resources are responsive to human interaction and these together with terrain feature determine the selection of proper land use pattern which reflects the cultural, social and economic conditions (Vink, 1975).

Basically, natural resources are directly or indirectly related to the surface cover in a given locality. The changes in any form of land use are largely related to the external forces and pressure built up within the system (Bisht and Kothyari, 2001). Impressions of growth on the ground are portrayed in the pattern of land use, structure and morphology. Thus, it is necessary to study the evolution of the land use pattern, structure and the layout. Transformation takes place on the land, over a period of time. The land use pattern of any area is a reflection, not only, of the immediate and current space requirements of the community but also rather of the cumulative needs over a period of years. Any effective planning for future should consider the present pattern of land use and factors responsible for its change.

Objectives

The study has been carried out with the following objectives:

- To understand geographical background of the study area.
- To generate base map for the study area.
- To classify land use/land cover for the study area using the satellite remote sensing data and supervised classification algorithm.
- To assess the change in land use/land cover in the study.

Study Area

Vallabhnagar tehsil of Udaipur district has been taken up as the study area for the project and is bounded by 24°19′37″ N to 24°44′24″ North latitudes and 73°51′59″E to 74°22′50″ East longitudes. Vallabhnagar tehsil is situated in the eastern part of Udaipur district flanked by tehsil Mavli in the north, by tehsil Girwa in the west, by tehsil Dhariyawad in the south and by tehsil Dungla of district Chittorgarh in the east covering an area of 1032.43 sq km. Vallabhnagar tehsil comprises of two sub-tehsils, i.e., Bhindar and Kanor 6 ILR circles, 48 Patwar circles, and 310 villages.

Land Use/Land Cover Analysis and Change Detection Studies

Information on the existing land use/land cover pattern, its spatial distribution and changes in the land use pattern is a prerequisite for planning, utilization and formulation of policies and programmes for making any developmental plan. This information not only provides a better understanding of land utilization aspects but also plays a vital role in developmental planning. Traditionally for regional planning exercises from macro-to micro-level, statistical data available under ninefold revenue classification of land use is used for studying and analysing the land use pattern at regional and intra-regional level. Spatial information analysis sometime is also supported by the departmental maps and conventional sources. With the advent of remote sensing techniques, both aerial and satellite, it has been possible to prepare land use maps at various levels showing categories of land under different uses both in spatial and statistical form. An attempt has been made to prepare maps of land use/land cover information with remote sensing data and to find out the changes in the land use pattern using a GIS package.

Land Use/Land Cover System in India

The land use/land cover system adopted by almost all concerned organizations and scientists, engineers and remote sensing community who are involved in mapping of earth surface features, is a system derived from the United State Geological Survey (USGS) land use/land cover classification system. This system was designed on the basis of the following criteria (Lillesand and Kiefer 1999):

(i) the minimum level of interpretation accuracy using remotely sensed data should be at least 85 per cent,

(ii) the accuracy of interpretation for the several categories should be about equal,

(iii) repeatable results should be obtainable from one interpreter to another and from one time of sensing to another,

(iv) the classification system should be applicable over extensive areas,

(v) the categorization should permit land use to be inferred from the land cover types,

(vi) the classification system should be suitable for use with remote sensor data obtained at different times of the year,

(vii) categories should be divisible into more detailed subcategories that can be obtained from large-scale imagery or ground surveys,

(viii) aggregation of categories must be possible,

(ix) comparison with future land use and land cover data should be possible, and

(x) multiple uses of land should be recognized when possible.

The basic USGS land use/land cover classification system for use with remote sensor data is shown in Table 11.1. On the basis of this system a multi-level system has been devised because different degrees of detail can be obtained from aerial and space images, which depend upon the resolution. In principle, levels IV and III are designed for local level or very large scale mapping, whereas level I and II and are meant for small scale mapping. Levels III and IV are logical extensions of levels I and II. Table 11.1 lists representative interpretation formats for various land use/land cover classification levels.

Land Use/Land Cover Classification System

The paper deals with the classification system that is being adopted for mapping, using satellite data. Land use refers to "man's activities and the various uses which are carried on land". Land cover refers to "natural vegetation, water bodies, rock/soil, artificial cover, resulting due to land transformations". Land use is generally inferred based on the land cover. Land use encompasses several aspects of man's relationships with the environment, e.g., buildings as a cover for residential use, green plants as a cover for agricultural crop. In some of the writings, it includes residential, commercial, industrial, recreational and institutional functions. In others, it signifies firstly, activity patterns of people in the urban setting and their institutions; and secondly, physical facilities or improvements to the land that are effected to accommodate these activities pattern. In some of the planning studies, the term, land use has been used not only in the sense of activity-use relationships but also in the value systems that the community plays in regulating space and its use patterns.

Classification Principles

Guttenberg developed a multiple land use classification system based on a set of classification principles and each based on a single criterion such as:

- General site development.
- Site adaptation (Building type).
- Actual use or activity.
- Economic function.

Table 11.1

USGS Land Use/Land Cover Classification System for Use with Remote Sensor Data

	Level I		*Level II*
1	Urban or Built-up Land	11	Residential
		12	Commercial and Services
		13	Industrial
		14	Transportation, Communications and Utilities
		15	Industrial and Commercial Complexes
		16	Mixed Urban or Built-up land
		17	Other Urban or Built-up land
2	Agriculture Land	21	Cropland and pasture
		22	Orchards, groves, vineyards, nurseries and ornamental horticultural areas
		23	Confined feeding operations
		24	Other agricultural land

Cont'd...

...Cont'd

3	Rangeland	31	Herbaceous Rangeland
		32	Shrub-Brush & Rangeland
		33	Mixed Rangeland
4	Forest Land	41	Deciduous Forest Land
		42	Evergreen Forest Land
		43	Mixed Forest Land
5	Water	51	Streams and Canals
		52	Lakes
		53	Reservoirs
		54	Bays and Estuaries
6	Wetland	61	Forested Wetland
		62	Non-forested Wetland
7	Barren Land	71	Dry Salt Flats
		72	Beaches
		73	Sandy Areas other than Beaches
		74	Bare Exposed Rock
		75	Strip Mines, Quarries and Gravel Pits
		76	Transitional Areas
		77	Mixed Barren land
8	Tundra	81	Scrub & Bush Tundra
		82	Herbaceous Tundra
		83	Bare Ground Tundra
		84	Wet Tundra
		85	Mixed Tundra
9	Perennial Snow or Ice	91	Perennial Snowfields
		92	Glaciers

The above system of land use classification gives flexibility and is expandable. The relationship among these individual dimensions of land use can also be of paramount importance to future planning. Polle has pointed out two approaches in developing land use classification for a specific area as:

Checklist approach: Various aspects are studied in the field and the classifications are formed.

Conceptual approach: The land use and land cover types are conceptualized and the activities in the field are related to one of the classified uses.

Classification Criteria

To develop any land use/land cover classification system, it is essential to consider certain criteria in order to arrange or group under a suitable framework to facilitate systematic land use inventory, mapping and change detection studies. It becomes essential to develop a standardized classification system, which is not only flexible in nomenclature and definition but also capable of incorporating information from remotely sensed data and other sources to satisfy user needs. For this, certain guidelines and criteria for land use/land cover classification system are being considered:

- The classification system should be applicable over a large area.
- Classification system should be suitable for using remotely sensed data obtained at different points of time.
- The minimum interpretation accuracy and reliability in the identification of land use categories for aerial data should be at least 80 to 90 per cent.
- The nomenclature, definition and framework to the extent possible should be compatible with existing terminology adopted by planning agencies.
- Classification should be easier to understand and flexible.
- Aggregation of similar or multiple land use classes should be possible at different levels of requirement.

Classification System Adopted

The classification system developed by aggregation and selection of various land use/land cover classes are mentioned below with its descriptions and operational definitions. Homogeneity of various phenomenon, their interrelationships and variations from their surroundings are being considered as prime criteria for identification and delineation of land use/land cover classes. The classes are:

- Degraded land under plantation cultivation.
- Degraded notified forest land.
- Land with or without scrubs.
- Miscellaneous polygons.
- Settlement.
- Water bodies.
- Mining and industrial waste.

Degraded Land Under Plantation Cultivation

Land except cultivated and fallow land, which is degraded and is used for plantation cultivation is included in this class.

Degraded Notified Forest Land

The land under natural vegetation, covered with reserved or protected forests is covered in this category.

Land with or without Scrubs

The land under natural vegetation, tree cover with good canopy, covered with or without scrubs, social forest plantations, etc., is covered in this category.

Miscellaneous Polygons

The class miscellaneous polygons include:

(i) *Cultivated land:* It covers all the cultivated land with or without crops including fallow land and agricultural farms are considered in this class. This land is used primarily for production of food, fibre and fodder.

(ii) *Wastelands:* This class includes lands presently lying unutilized due to various constraints. It also includes derelict lands, which are unutilized, partially utilized or mismanaged. Wastelands generally include gullied/eroded lands which are the first stage of land dissection followed by their networking leading to creation of ravines and waterlogged areas which include the land where water is at or near the surface and water generally stands for a considerable time in a year excluding the natural or man-made water bodies.

(iii) *Vacant lands:* Vacant land is a non built-up land without any land use activity or land cover. These are mainly barren lands devoid of trees/bushes/scrub.

Settlements

Settlements are spotted all over the study area. Some compact clusters of dwelling units are found outside the main urban built-up areas surrounded by agricultural land. These residential areas are generally unplanned and characterized by irregular layout.

Water Bodies

The land covered under natural drainage system like rivers, streams and natural drains as well as man-made linear drainage system like canals (used or unused) and natural or man-made structures like reservoir or ponds is classified under water bodies. It has following three subgroups:

(i) River/Drains
(ii) Lakes/Ponds
(iii) Canals

Mining and Industrial Waste

This category contains lands having mining and industrial waste.

Remote Sensing Techniques for Land Use/Land Cover Analysis

Remote sensing techniques are used for mapping existing land use as well as constant monitoring of changes in land use/land cover pattern. Remote sensing data products providing almost real time information are not only valuable tools but also cost and time effective. Remote sensing that has the most advanced method of data acquisition system can provide accurate, timely, reliable and up-to-date spatial information at a regular interval. At present it is found that remote sensing techniques integrated with GIS tools have capabilities to generate, process and analyse the spatial database more efficiently and accurately in a comparatively less time and at a faster rate.

Image Classification

Image classification is a procedure to automatically categorize all pixels in an image of a terrain into land cover classes. Normally, multi-spectral data are used to perform the classification of the spectral pattern present within the data, for each pixel is used as the numerical basis for categorization. This concept is dealt under the broad subject, namely, 'Pattern Recognition'. Spectral pattern recognition refers to the family of classification procedures that utilizes this pixel-by-pixel spectral information as the basis for automated land cover classification. Spatial pattern recognition involves the categorization of image pixels on the basis of the spatial relationship with pixels surrounding them. Image classification techniques are grouped into two types, namely supervised and unsupervised. The classification process may also include features, such as, land surface elevation and the soil type that are not derived from the image. A pattern is thus a set of measurements on the chosen features for the individual to be classified. The classification process may therefore be considered as a form of pattern recognition, that is, the identification of the pattern associated with each pixel position in an image in terms of the characteristics of the objects or on the earth's surface.

Supervised Classification

A supervised classification algorithm requires a training sample for each class, that is, a collection of data points known to have come from the class of interest. The classification is thus based on how "close" a point to be classified is to each training sample. We shall not attempt to define the word "close" other than to say that both geometric and statistical distance measures are used in practical pattern recognition algorithms. The training samples are representative of the known classes of interest to the analyst. Classification methods that rely on use of training patterns are called supervised classification methods. The three basic steps involved in a typical supervised classification procedure are as follows:

(i) *Training Stage:* The analyst examines representative training areas and develops numerical descriptions of the spectral signatures of each land cover type of interest in the scene.

(ii) *The Classification Stage:* Each pixel in the image data set is categorized into the land cover class it most closely resembles. If the pixel is insufficiently similar to any training data set it is usually labelled 'Unknown'.

(iii) *The Output Stage:* The results may be used in a number of different ways. Three typical forms of output products are thematic maps, tables and digital data files which become input data for GIS for spatial analysis of the terrain and depicts the flow of operations to be performed during image classification of remotely sensed data of an area, which ultimately leads to creating a database as an input for GIS.

There are a number of powerful supervised classifiers based on the statistics, which are commonly used for various applications. A few of them are a minimum distance to means method, average distance method, parallelepiped method, maximum likelihood method, modified maximum likelihood method, Baysian's method, decision tree classification, and discriminate functions.

Unsupervised Classification

Unsupervised classification algorithms do not compare points to be classified with training data. Rather, unsupervised algorithms examine a large number of unknown data vectors and divide them into classes based on properties inherent to the data themselves. The classes that result, stem from differences observed in the data. In particular, use is made of the notion that data vectors within a class should be in some sense mutually close together in the measurement space, whereas data vectors in different classes should be comparatively well separated. If the components of the data vectors represent the responses in different spectral classes, as opposed to information classes, they represent the ground cover types of interest to the analyst.

Change Detection Studies

Change detection involves the uses of different data set to discriminate areas of land cover changes between dates of imaging. The types of changes that might be of interest can range from short-term phenomena such as snow cover or floodwater to long-term phenomena such as urban fringe development or desertification. Ideally, change detection procedure should involve data acquired by the same (or similar) sensor and be recorded using the same spatial resolution, viewing geometry, spectral bands, radiometric resolution, and time of day but conditions other than this can also be used for change detection studies.

The change detection process may also be strongly influenced by various environmental factors that might change between image dates. In addition to atmospheric effects, such factors like level, tidal stage, wind, or soil moisture condition might also be important.

One way of discriminating changes between two dates of imaging is to employ post-classification comparison. In this approach two dates of imagery are independently classified and registered. Then an algorithm can be employed to determine those pixels with a change in classification between dates. In addition, statistics (and change maps) can be compiled to express the specific nature of the changes between the dates of imagery. The accuracy of such procedures depends upon the accuracy of each of the independent classifications used in the analysis. The errors present in each of the initial classifications are compounded in the change detection procedure.

Another approach to change detection using spectral pattern recognition is simply the classification of multi-temporal data sets. Supervised or unsupervised classification is used to categorize the land cover classes in the combined image. The success of such efforts depends upon the extent to which "change classes" are significantly different spectrally from the "non-change" classes.

Principal components analysis is sometimes used to analyse multi-date image composites for change detection processes. In this approach, two (or more) images are registered to form a new multi-band image containing various bands from each date.

Temporal image differencing is yet another common approach to change detection. In the image differencing procedure, DNs from one date are simply subtracted from those of the other. The difference in areas of no change will be very small (approaching zero), and areas of change will manifest larger negative or positive values.

Methodology and Approach

Image classification is a procedure to automatically categorize all pixels in an image of a terrain into land cover classes. Normally, multi-spectral data are used to perform the classification of the spectral pattern present within the data, as each pixel is used as the numerical basis for categorization.

The digital processing of satellite images are used for land use/land cover analysis and change detection studies. The satellite images of IRS 1C/1D (LISS III) data are used in the present study for land use/land cover analysis of Vallabhnagar tehsil. The various thematic layers are prepared using the Survey of India toposheets. The whole study is carried out in following stages:

(i) Data collection and interpretation/mapping.
(ii) Data verification.
(iii) Map finalization/database creation.
(iv) Data analysis.

Selection of Parameters for Land Use/Land Cover

To develop any land use/land cover classification system, it is essential to consider certain criteria in order to arrange under a suitable framework to facilitate systematic land use inventory, mapping and change detection studies. The important features used for analysis are:

(i) Agriculture land.
(ii) Natural vegetation.
(iii) Settlements.
(iv) Bare land.
(v) Dense scrub land.
(vi) Water body.

Satellite Data Used

- Satellite data of IRS-LISS-III, January 1999.
- Satellite data of IRS-LISS-III, January 2004.

- Survey of India (SOI) Topographical sheets on 1:50,000 scale for preparing different thematic layers.
- Ground truth data for doubtful areas.

Geo-referencing of Satellite Data

The satellite data of both IRS are geo-referenced to bring them to the same coordinate system. The both imageries are in Lambert Conformal Conic projection with modified Everest as spheroid.

Base Map Preparation

The base map is referred to a map containing the basic reference details of infrastructure information such as road, railway, water body, etc., which act as points of reference while transferring the thematic details from topographical map on 1:50,000 scale. The base map details are further updated using satellite image.

Ground Truth Verification

The ground truth verification is done for land use/land cover analysis for the doubtful areas. Ground truth verification of doubtful areas enhances the interpretation accuracy. The doubtful areas were physically verified in the field, and field observation about the terrain conditions and land use patterns were noted for use in the modification of thematic details.

Supervised Classification

In supervised classification, the analyst makes use of prior knowledge available from topographic, photo-interpreted maps and field surveys while training the computer by way of assigning "training areas" about different features that are to be classified by the computer. Signature is grouped and assigned to respective classes.

Various supervised classification methods are used to assign an unknown pixel to one of a number of classes. Among them, the most frequently used classification algorithms are:

- Minimum distance to mean.
- Parallelepiped classification.
- Maximum likelihood algorithm.

Land Use/Land Cover Change Detection

Statistics of land use changes was done in GIS environment. For change detection, the data of IRS-LISS-III were used. Classification of individual data sets and pixel-to-pixel comparison through intersection Matrix analysis has been carried out to assess the change in land use/land cover between the years 1999 and 2004.

The total number of classes is varying from area to area. The imagery has to be interpreted and ground checked for corrections.

Result and Discussion

The changes occur in land use/land cover pattern of any area with time. This may be due to man-made activities on the land or due to natural causes. The study of temporal change of land use/land cover over fixed time interval is necessary for the proper management of any land.

The present study depicts the changes in the land use/land cover pattern of Vallabhnagar tehsil area between the years 1999–2004. The highlights of the change detection study map are shown in Table 11.1.

The changes in the land use/land cover can be said to be positive change if wastelands have been converted to usable land, such as crop land. On the other hand, the changes are said to be negative if the crop land has degraded to wastelands.

The change detection has been carried out by analysis of IRS LISS III data acquired in January 1999 and January 2004. The highlights of the change detection are given in Table 11.2.

Remote sensing in combination with GIS is very useful for the preparation of land use/land cover maps of an area. A digital classification technique has been carried out for the study area of Vallabhnagar tehsil. IRS LISS – III data set has been used and supervised classification techniques has been carried out for digital classification.

The change detection studies have been carried out for Vallabhnagar tehsil to determine the change in land use/land cover over a period of time. The change has been assessed between the years 1999 and 2004. The result obtained for the land use/cover analysis and the change detection studies of the study area are presented and discussed in this chapter.

Land Use/Land Cover Analysis of Vallabhnagar Tehsil

The land use and land cover categories such as settlements, agriculture, natural vegetation, water bodies, forest land and dense scrub land have been identified using supervised classification techniques. The land cover statistics are given in Table 11.2.

Table 11.2

Land Use/Land Cover Statistics for Vallabhnagar Tehsil (as per IRS Satellite Data, January, 1999 & 2004)

S. No.	Land use	Area in Hectares		Percentage	
		1999	2004	1999	2004
1	Degraded land under plantation crops	125	0	0.1%	0.0%
2	Degraded forest	111	1,310	0.1%	1.3%
3	Land with or without scrubs	38,612	29,476	37.4%	28.6%
4	Cultivated agricultural land	58,853	68,302	57.1%	66.2%
5	Settlements	1,198	1,779	1.2%	1.7%
6	Water bodies	4,244	1,582	4.1%	1.5%
7	Mining/Industrial waste	0	694	0.0%	0.7%
	Total	10,31,143	1,03,143	100.0%	100.0%

The land use/land cover pattern of Vallabhnagar tehsil is discussed below:

(i) Degraded land under plantation cultivation: Degraded land under plantation cultivation was 125 hectares in 1999, which merged in other classes of land in 2004. Hence, there was no degraded land under plantation cultivation in 2004.

(ii) Degraded notified forest land: Degraded notified forest land was 111 hectares in 1999 and 1,310 hectares in 2004. Hence, the area of degraded notified forest land had increased.

(iii) Land with or without scrubs: Land with or without scrubs was 38,612 hectares in 1999 and 29,476 hectares in 2004. Hence, the area of land with or without scrubs had decreased, as the land was put to more agricultural purposes.

(iv) Cultivated agricultural land: From the satellite data the agriculture area can be clearly delineated. On the aggregate classified output of agriculture land was 58,853 hectares in 1999 and 68,302 hectares in 2004. Hence, the area under cultivated agricultural land had increased, as the land was put to more agricultural purposes.

(v) Settlement: It is defined as an area of human habitation. The area estimated under this category was 1,198 hectares in 1999 and 1,779 hectares in 2004. Hence, the area under settlement had increased, owing to growth of population.

(vi) Water body: Ponds, lakes, rivers and streams are the important water bodies seen in the study area. The area statistics for this category was 4,244 hectares in 1999 and 1,582 hectares in 2004. Hence, the area under water bodies had decreased, owing to scarcity of rainfall.

(vii) Mining and industrial waste: There was no land used for mining and industrial Waste in 1999. However, about 694 hectares land was used in 2004 for mining and industrial waste. This was due to a cement factory that came into existence in Daroli village.

Plate 11.1

Land Use/Land Cover Map Vallabhnagar : Year 1999

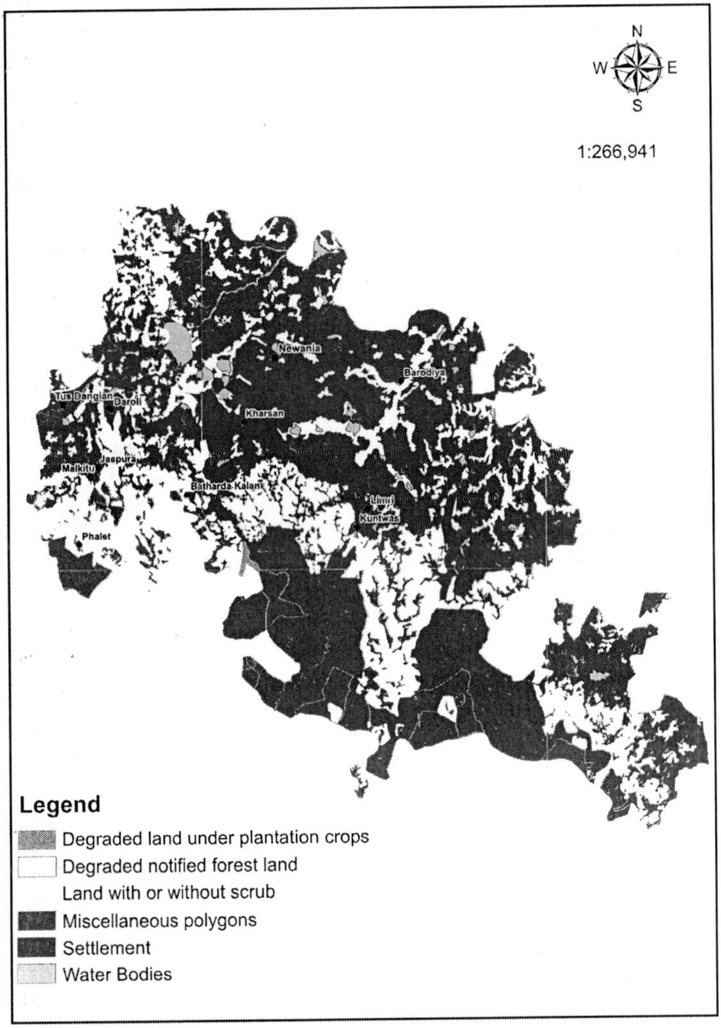

The major land cover of the study area is agricultural land covering 57.1 per cent in 1999 and 66.2 per cent in 2004 of the whole geographical area.

The land use/land cover pattern of Vallabhnagar tehsil is shown in Plates 11.1 and 11.2.

Plate 11.2

Land Use/Land Cover Map Vallabhnagar : Year 2004

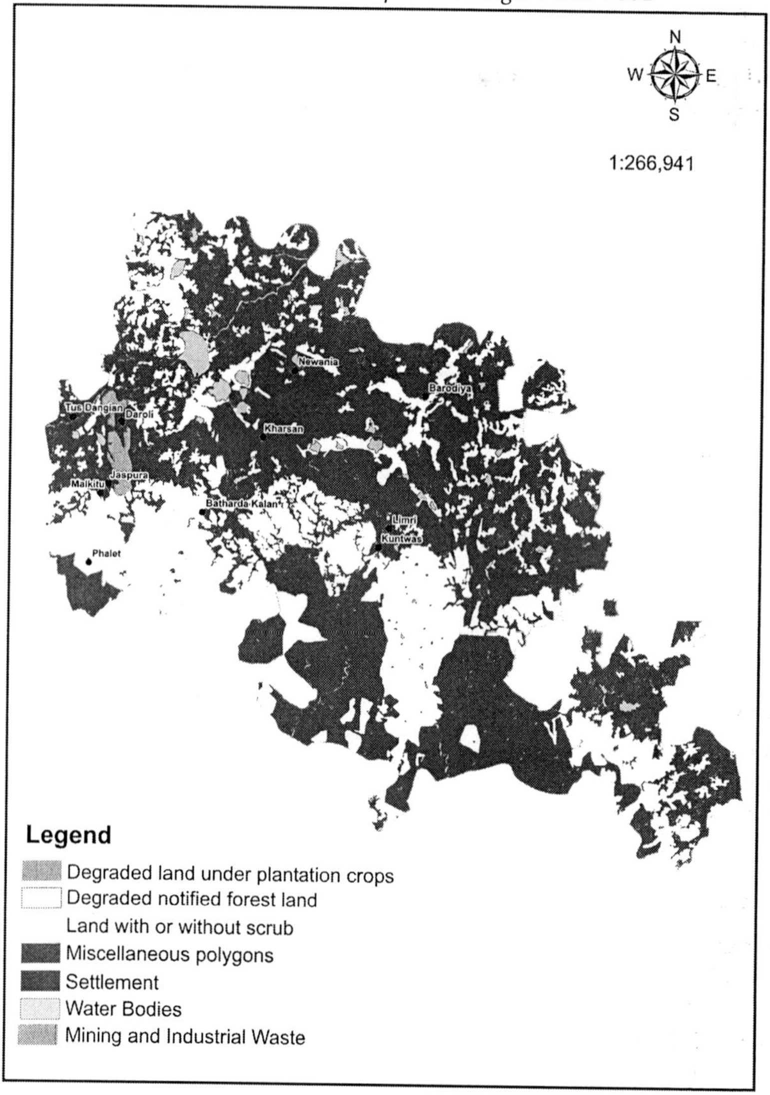

Conclusion

The quantification of various land use/land cover categories and its distribution was successfully done with the help of remote sensing and geographical information system. The remote sensing technique is found to be faster, accurate and useful for mapping or generating a spatial database for an area. The GIS tools are used in deriving desired or user-required information from the database along with statistical analysis. Because of the distinct advantages in terms of the timeliness, repeativity, real time capability of data reception and transmission, and computer compatibility, remote sensing techniques are indispensable tools for land use/land cover mapping and trend analysis. GIS has proved to be a powerful tool for integrating these various information and generation of the geo-database.

The aim of this study was to interpret the land use/land cover features using satellite remote sensing digital techniques and GIS. In addition, the change in the land use/land cover pattern between 1999 and 2004 has been assessed through change detection studies.

The changes were observed by comparing the two supervised classifications and thematic layers of the study area. The area under degraded forests, cultivated agricultural land, settlements and mining and industrial waste has increased, whereas area under degraded land under plantation crops, land with or without scrubs and water bodies has decreased as compared with 1999 and 2004 data.

From the above discussion, it is established that by the help of remote sensing and GIS the analysis of existing land use of the area can be effectively determined and can be used for future planning purpose.

Acknowledgement

The paper is based on project work carried out by Dr R.P. Sharma, under supervision of Prof. P.R. Vyas (Internal Supervisor) and Mr Sudheer Bhatnagar, Geologist (R.S.), Directorate of Mines and Geology, Udaipur.

References

Anderson, J.R. (1972), 'Land Use/Land Cover Changes: A Framework for Monitoring', *Journal of Research*, Vol. 5, No. 3, U.S. Geological Survey, Government of India, pp. 143–53.

Anuta, P.E. (1977), 'Computer Assisted Analysis Techniques for Remote Sensing Data Interpretation', *Geophysics*, Vol. 42, pp. 468–81.

Aronoff, S. (1982), 'Classification Accuracy: A User Approach', *Photogrammetric Engineering and Remote Sensing*, Vol. 48.

Bisht, B.S. and B.P. Kothyari (2001), 'Land-Cover Change Analysis of Garur Ganga Watershed Using GIS/Remote Sensing Technique', *J. Indian Soc. Remote Sensing*, Vol. 29, No. 3, pp. 165–74.

Campbell, J.B. (1983), *Mapping the Land Aerial Imagery for Land Use Information*, Association of American Geographers, Washington, DC.

Drury, S.A. (1990), *A Guide to Remote Sensing*, Oxford University Press, Oxford, p. 199.

Jensen, J.R. (1983), 'Urban-Suburban Land Use Analysis', Chapter 30 in R. N. Colwell (ed.), *The Manual of Remote Sensing*, Vol. 2, American Society of Photogrammetry and Remote Sensing, Falls Church, pp. 1571–966.

Lindgren, David T. (1985), *Land Use Planning and Remote Sensing*, Martinus Nijhoff Publishers, Boston, pp. 101–15.

Pidwirny, M. (2006), 'Introduction to Physical Geographic Information System', *Fundamentals of Physical Geography*, 2nd Edition.

Rao, D.P. (1995), *Remote Sensing for Earth Resources*, Association of Exploration Geophysicists, Osmania University, Hyderabad, pp. 153–69.

Sokhi, B.S. (1993), 'Remote Sensing in the Study of Compatibility of Land Uses and a High Capacity Intra Urban Transportation System, New Delhi', *Asian Pacific Remote Sensing Journal*, Vol. 6, No. 1.

Uttarwar, P.S., N.D. Sharma and B.S. Sokhi (1989), 'Satellite Remote Sensing in Urban Sprawl Mapping and Monitoring: A Case Study of Delhi', *Journal of the Indian Society of Remote Sensing*, Vol. 17, No. 3.

Vink, A.P.A. (1975), *Land Use in Advancing Agriculture*, Springer-Verlag, Berlin, p. 394.

Web Link References

GOES 3.9 um Channel Tutorial – An Excellent Tutorial on Thermal Remote Sensing.

http://www.physicalgeography.net/fundamentals/2e.html S.A.

Implementing GIS in Lebanon – A Case Study by Jacques Ekmekji, Director GIS Services Division/Associate Khatib and Alami – Consolidated Engineering Company Beirut Lebanon.

Indian Remote Sensing Satellites, History of Indian RS Satellites – An Article by Wim Bakker – ITC.

Remote Sensing of the Global Environment – An Article by David J. Schneider, Michigan Technological University.

12

Land Use Planning of Vaniyambadi Block, Vellore District, Tamil Nadu

Rahul Tiwari and Pradeep Purohit

Land resource is the mother of all natural resources, which provides life support system to all living beings. The water resource: streams, lakes, and groundwater is also a product of the land. Land and water support the plant and animal kingdom. Conservation, development and management of these resources are the key to food and employment, security, biomass supply, healthy environment and social stability. Unfortunately, human and animal population are rising, but per capita landholding is declining.

The resources are degrading due to soil erosion, salinization, waterlogging, and unscientific management practices, etc., sapping its productive capacity, posing a threat to the security of food and livelihood. Due to the continuously increasing human and animal population the water table is going down at an alarming rate. So it becomes the duty of planners, scientists, technologists and soil conservationists to develop and pursue vigorously relevant science and technology for managing the precious land and associated resources to ensure sustainable productivity.

Land is an age-old precious gift of nature to the mankind and it takes 600–1000 years to make one-inch top layer of soil. This precious resource of nature can be washed away by a single heavy storm of rain. It is estimated that soil is losing its fertility at the rate of 5 acres/minute. This resource is degrading due to soil erosion, mismanagement practices, etc. The land use study is of fundamental significance, as the land resources play a strategic role in the determination of man's economic, social and cultural progress. In fact, the land use of a region is always characterized by the spatial variations and is profoundly influenced by physio-socio economic factors. The last two centuries have witnessed the scientific and technological developments which have brought about tremendous change in land use.

Study Area

Vaniyambadi is a taluka bounded by Gudiyattam Block in the north, Vellore Block in the east, Andhra Pradesh State on the west and Tirupattur Block on the south directions. The study area is situated in the north-west part of the Vellore district which lies between latitude 12°68′N and longitude 78°62′E. The total geographical area is 84,007 hectares.

Objectives

The objectives of the study are:

(i) To evaluate the land resources of the block using satellite data.
(ii) To create and organize an error-free digital data for all the natural resources in the form of separate layers for the study area.
(iii) To analyse the land resource data and integrate it in GIS domain.
(iv) To generate optimal land use plan in GIS domain.

Methodology

Land Use Planning Introduction

Land use planning in rural areas has received scant attention in comparison with town planning, until recently. This can hardly be justified in a country like India, which is predominately agricultural. The planning of agricultural sector in rural areas has been carried out in a laissez-faire manner; very few deliberate cautious attempts were made to improve the countryside. The concept of land-use planning is not recent in its origin but its practice is truly recent. A land use map gives us thorough and clear picture of land to the planners for the determination of future use and planning the agricultural sector to maintain the land potentials. In this, wetland-use planning is concerned with the future use of land and the changing demands of the society.

The advent of computers and their ever-increasing power of handling large data sets and computations have provided us with tools to analyse data and obtain results within lesser time than was required earlier. The availability of satellite data for large areas to assess the natural resources and reconnaissance surveys within a short period has forced us to use the information of planning and development. The GIS as a tool to spatially correlate information and look at it in a new perspective has helped in decision-making at the right time. Within the availability of the above, the present study has successfully proved the validity of the adoption of the new technology in the planning and development.

The use of remotely sensed data for such study is appropriate to the extent that it is the most suited technology to be used for a country as large as India and its resources study. The first task in this study was to take stock of land resources, its use and misuse and finally suggest alternate land use, if necessary at all and/or land and water management practices for sustainable availability of resources. The first task in carrying out an integrated land use survey is to undertake an assessment of the land use pattern as a basis of subsequent field studies. This will indicate land capability, after which a land development plan indicating the best land use pattern for the area covered and taking into account all the relevant factors can be prepared.

An accurate assessment of the existing land use pattern is best carried out from an appraisal of complete vertical photography of the project area or satellite data around 1:25000. Such an initial appraisal giving the overall land use pattern enables the specialist concerned to see the framework on which they will have to work.

Land Use/Land Cover of the Block

Land use/land cover categories of the block have been grouped into eight levels – built-up area, crop land, fallow land, open scrub, river, scrub, settlement, tank, etc.

Built-up Land

These are the areas of human habitation development due to intensive non-agricultural use. The major town is Vanlyambadi of the block. It occupies 0.16 per cent of the total area.

Crop Land

Crop land is primarily used for production of food, fibre, commercial and horticultural crops. This category is largely dependent on agro-climatic conditions and it is dominant category in the study area. It occupies 46.86 per cent of total area.

Table 12.1

Distribution of Crop Land Area (Hect.)

Rice	*Chadam*	*Cumbu*	*Ragi*	*Sugar Cane*	*Pulses*
6,922	1,567	344	2,082	4,320	1,683
Cotton	Ground Nut	Banana	Coconut	Others	Total Area (Hect.)
4,605	9,154	1,942	943	13,571	47,133

Fallow Land

Fallow land is temporarily allowed to remain uncropped during the agricultural year. This category could be delineated using temporal satellite data wherein the signatures of crops in the cropped area conspicuous by their absence as one of the data of satellite overpass. It occupies 0.10 per cent of total area.

Forest

Forest areas are associated with trees and other natural vegetation types. It occupies 21.68 per cent of the total area.

Barren Land

Barren land characterized by expensed massive rocks. It occupies 2.70 per cent of the total area.

Land put to Non-agricultural Use

Land put to non-agriculture use occupies 0.652 per cent of the total area of the block.

Cultivable Waste

Cultivable waste occupies 1.75 per cent of the total area of the block.

Permanent Pasture and Grazing Land

Permanent pasture and grazing land occupies 1.55 per cent of the total area of the block.

Miscellaneous Tree Crops

There are miscellaneous tree crops covering 0.29 per cent of the total area of the block.

Current Fallow

Current fallow occupies 15.39 per cent of the total area of the block.

Soil

Soil is the basic and non-renewable natural resource support for agriculture and various developmental activities and economic growth of a region. More so, it is true for a drought-prone area wherein the ecosystem is fragile and the sustainable development requires a very careful and optimal utilization of available land and water resources. While maintaining a good harmony with the environment for optimal utilization of available land and water resources, information on soil with respect to nature, physico-chemical characteristics, potential and inherent limitations is a prerequisite. Conventional soil surveys provide this information. Such approach, however, is not only tedious and time consuming but also at the same time cost-prohibitive. In early sixties, the development of aerial photo-interpretation technique and availability of aerial photographs opened a new vista in soil resources mapping programme in our country. Subsequently, the availability of space borne multi-spectral data especially from the Indian remote sensing series of satellites further augmented this process and has enabled soil scientists generating timely and cost-effective information on soil resources. The IRS-1C/1D multi-spectral and panchromatic stereo data with improved spatial resolution will further enhance this capability.

Generation of TIN and DEM

In order to establish flow accumulation points and possible stream network an elevation raster has been created. For this contour data thus generated in the vector format has been used to generate TIN (a surface processing data) using Arc Map 3D analyst functions. Later this TIN has been converted to Digital Elevation Model (DEM) raster form by spatial analyst functions. Slope map has been generated from the elevation raster by surface analysis tools under spatial analyst functions. Similarly, aspect map has also been prepared from the slope map by the same method.

Slope Map

Slope plays a crucial role, in addition, to find out the flow direction and flow accumulation in hydrological modelling. Slope is one of the important terrain parameters which is explained by horizontal spacing of the contours. In general, in the vector form closely spaced contours represent steeper slopes and sparse contours exhibit gentle slope, whereas in the elevation output raster every cell has a slope value.

Aspect

Aspect is the direction of slope with respect to north. In output slope raster generation, aspect identifies the down slope direction from each cell to neighbouring cells. The aspect values of the output raster will be measured in the compass direction, i.e., from N–NE–E–E–S–SW–W–NW (i.e., 0° – 360° clockwise).

Results and Discussion

Preparation of Action Plan

The strength of GIS is the integration of multi-layered data from different source and various scales. The integration of various layers of information has been a difficult task manually until the maps were drawn on a transparent film. With the availability of GIS, which takes the data into digital space, the ability to see through maps, which are overlapped one over the other digitally, analysis of the map is achieved. Database management systems integrated with graphic interface has a powerful query capability. This finally gives the analytical ability to pose complex queries and extract information spatially.

Remote sensing data is the final solution for large area coverage and temporal coverage of land area. It is an important source of input to GIS; different layers of information generated from remotely sensed data provides an authentic input when used in GIS. The analysis carried out in the present study used primarily information generated from remotely sensed data. Layers were generated for land use, drainage, soil, slope, TIN DEM, water holding capacity along with base map. The soil map is taken as the base. All the layers were brought to a common coordinate system so that integration is possible and the accuracy of output is maintained.

The preparation of action plan for land resources involves the integration of layers such as land use/land cover and soil. Arc/map GIS software is used for integration of layers. The layers were integrated using the right command. Once the layer is integrated, a multi-criteria approach has been adopted to obtain information for different combinations of land use and soil. Presentation of action plan involves the suggestion of alteration of present system of land use or alternate system depending on the local condition of prevalence of practice and socio-economic conditions.

Land Use Planning

Based on careful integration on information of soil, land use, slope, water holding capacity, the following action plan have been formulated in development of land, soil, water conservation.

Different themes on land resources of the study area were integrated and different landscape units were assessed. The scheme of integration were given in the table. A resource-based action plan was evolved for development conservation and management of the area on a sustainable basis. The different mapping units and the suggested action plan will therefore aid planners as well as other executing agencies in implementing various developmental programmes in the block.

Soil and Water Conservation Measures

Soil and water are the basic resources essential for survival of mankind on earth. Ironically, very few people realize the importance of conserving and judiciously utilizing the soil, the greatest gift of nature. According to H.H. Bennet, "Soil without water is desert and water without soil is useless". In fact, every kind of farm activity is connected with land and prosperity of a nation depends on the quality of its land resources. Left to nature, the soil is protected by inborn vegetation. Under undisturbed conditions, geological erosion of 'degradation' and 'aggradation' is well balanced: thus, a state of equilibrium is reached between climate, soil, rainfall, land slope and vegetal cover.

It is only by man's interference that the protective shield of land is disturbed and it is torn into pieces by the erosion process – starting from splash erosion to formation of gullies and ravines. Unprecedented expansion in man's needs is at the root of felling the trees and removal of grasses. As a result, erosion problem has assumed threatening dimensions.

According to the National Commission on Agriculture (1975) an estimated 50 million hectare out of the total geographic area of 328 million hectare are subjected to serious water and wind erosion; and 32 million hectare are prone to wind erosion. The area affected by gullies and ravines accounts for about 4 million hectare while the area affected by shifting cultivation is about 3 million hectare. There is 70 million hectare area under rainfed farming in the country.

Because of fragile nature of the rainfed geo-system, soil and water conservation programme is of paramount importance. Of late, experiences with different watersheds models have also suggested initiation of solutions to this serious problem.

Bunding

In agricultural lands, contour bunding is the most widely practised soil conservation measure for mildly sloping lands. Contour bunds are designed as mini-earthen dams with emphasis on height of bund and cross-sectional areas, depending on texture and depth of the soil. The contour cultivation – a supporting practice, is effective on land with very gentle slopes. This not only reduces the soil loss but also helps in increasing soil productivity.

Bunds (small earthen barriers) are provided in agricultural lands with slopes ranging from 1 to 6 per cent. They control the effective length of slope and thereby reduce the gain in velocity of run-off flow to avoid rill and gully formations. Important objectives of bunding are:

(i) To increase the time of concentration of rain water where it falls and thereby allowing more opportune time for rain water to be absorbed in the soil profile.

(ii) Converting a long slope into several short ones so as to minimize velocity and thereby reducing the erosive power of run-off water.

(iii) To provide field-to-field access for man and animals for undertaking agronomic operations.

(iv) To divert run-off water either for water harvesting purposes or for saving lower lands from excessive sand deposition or getting it severely eroded.

Farm Ponds

Farm ponds are bodies of water, made either by constructing an embankment across a watercourse or by excavating a pit or the combination of both. Basic aim of traditional farm ponds is:

- to provide water storage for life saving irrigation in a limited area.
- to provide drinking water for live stock and human beings in arid areas.
- to serve as water storage for providing critical irrigation to limited number of fruit plants for establishments.
- to moderate the hydrology of small watersheds.

Specific Site Conditions: Dugout ponds are generally created by excavating pits in areas having flat topography and mostly in situations where water table is close to the ground level. On the other hand, impounding type farm ponds are a common feature wherever there are well-defined waterways with rolling type of topography. It is possible to achieve high water storage to earth work ratios with embankments/bunds constructed in narrow and deep valleys, while the ponds constructed by excavation only, have this ratio mostly less than unity. In the case of impounding type, preference is given to locations with low soil permeability.

Broad Bed Furrow Trenching: Adoption of broad bed and furrow system, a land management technology developed by International Crops Research Institute for the Semi Arid Tropics, (ICRISAT), Hyderabad for deep black soils is recommended for these soils. It is a system of semi-permanent broad beds and furrows that are laid out on a gradual slope – usually 0.4–0.8 per cent. Each bed approximately, 100 cm wide, is slightly raised, acting as an 'in situ' bund for good moisture conservation and erosion control. The furrow (about 50 cm in width) is shallow (about 15 cm) and provides good surface drainage to prevent the waterlogging of crops grown on the bed. Excess water is let off through a system of field drains. Some land smoothing is usually necessary to remove small depressions and high spots. Another feature of broad bed and furrow system is that the run-off water can be harvested in a tank, which subsequently could be recycled for supplemental irrigation.

Once the drainage conditions improve, these areas become suitable for intensive agriculture like sequential cropping in kharif and rabi seasons or inter-cropping during kharif season. A wide range of cropping systems is available to be adopted in the district. Some of the recommended systems are jowar, maize, soybean, green gram, black gram followed by wheat, chick pea, mustard, lentil and linseed. Red gram can also be inter-cropped with the kharif crops.

Vegetative Barriers

Vegetative barriers are closely spaced grass hedges or plantations, usually a few rows of grasses or shrubs – grown along contours or with little grade for erosion control in agricultural lands. Thus, practices are gaining ground of vegetative barriers (e.g., Vetiver hedgerows or leucaena, lemongrass, *Cenchrus ciliaris*) alone at suitable vertical interval may be sufficient for run-off and erosion control in relatively flat and slightly undulating topography. But it is safer to have vegetative barriers only as inter-terrace treatments.

Specific Site Conditions

Vegetative barriers can be easily established across a wide spectrum of soil-climatic conditions except in class VII lands and desert conditions. Selection of species depends upon site-specific conditions, particularly soil and climatic variables. The major constraint experienced in their sustenance is stray cattle grazing.

Grass Waterways

Grass waterways are drainage channels either developed by shaping the existing drainageways or constructed separately for effecting drainage of agricultural lands. They are aligned along the major slope to handle run-off discharge from contour/graded bunds, bench terraces, contour trenches and contour furrows.

Specific Site Conditions

As far as possible, waterways are located along valley lines. But, sometimes it may be necessary to construct waterways along field boundaries for safe disposal of excess rainfall from agricultural fields. Waterways may be located in all classes of lands except bare rocks, where construction may be difficult. However, vegetative waterways should not be used for handling continuous flows, like that from tile drains, as problem of wetness may result in poor vegetal growth and soil protection.

Gully Control Measures

Depending on the severity of erosion, different types of mechanical protection measures have been recommended for arresting soil erosion and surface water run-off.

Contour Trenching

The trenches may be staggered or of continuous type. Their spacing or vertical interval depends on the slope of the land. Spacing may vary from 30–60 metres. Normally, continuous trenches are made with a cross section not exceeding 60 cm × 30 cm × 60 cm. The cross-section is designed to collect and convey the run-off expected from spaces between two successive trenches. Staggered trenches are of shorter length from 4–15 m with a cross section of 30 cm × 30 cm. In the alternative row, the trenches are located directly below one another. On the excavated soil piled on the downhill slope, grass seeds may be sown and saplings may be planted.

Check Dams

These structural measures are desirable for medium and large-sized gullies. Small gullies (depth < 1.0 m) can be controlled by improvement of gully catchments and diversion of run-off above the gully area. Depending on the site characteristics and availability of low cost material, either temporary or permanent check dams can be constructed. Temporary check dams – which include earthen, brushwood, loose stones or boulders, stone slabs, etc., are able to retain major portion of the run-off discharged from drainage area within specified period. However, a series of check dams will be more effective in erosion control, where multiple gully system exists.

The water resources development and management influence to a great extent, the land resource management. Therefore, it is necessary to develop, conserve and efficiently utilize the available surface water and groundwater. To this end, excess rainwater needs to be conserved/stored in different storage structures which could be directed to artificially recharge the groundwater for its use later.

Conclusion

Earth observation satellites provide the vantage point and coverage necessary for studying our planet as an integrated physical and biological system. Over the past decade, India has continuously and successfully shared the platform with the developed countries in the monitoring of the earth with its own series of satellites. The IRS series satellites have not only fulfilled the national mission in integrated development but are also a source of international standard data. The present study has utilized the IRS – 1 B data set. The data set not only satisfies the needs of the planning, at which the study is done but also provides a fastest best solution for a large area comprehensive survey.

Plate 12.1

Land Use/Land Cover of Vaniyambadi Block of Vellore

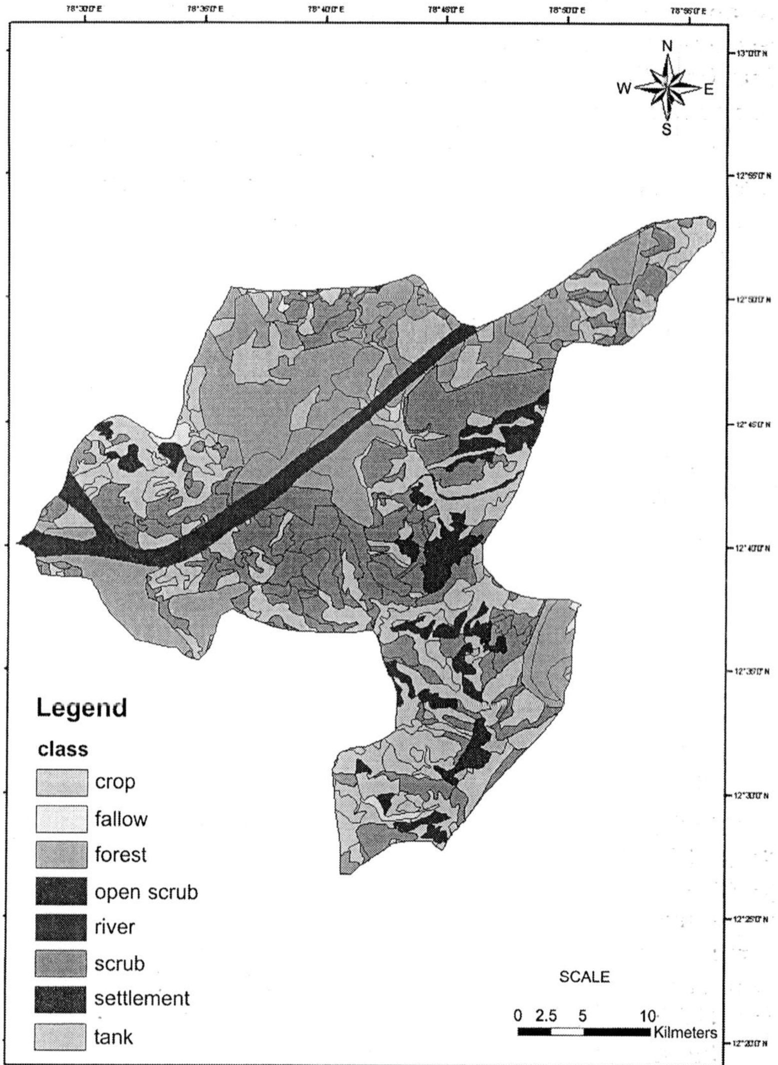

Land use planning involves the inventory of the land resources and taking stock of the present scenario. This is the first task to be taken by the planners towards attaining the goal. This inventory shows the land use pattern in the Drought Prone Area Project (DPAP) blocks identified by the Government of India for a prompt action. For such a type of study, the task of inventorying the natural resources is to be done within a set time-frame. The remotely sensed data from the IRS satellite is the ideal choice for its technical and economical viability.

The IRS-1 BLISS II sensor with spatial resolution of 32.51 m is used to obtain information on land use over two seasons. Soil mapping is also carried out in the summer data set. Landform mapping for water resources targeting is also carried out using the IRS-1B data. With sufficient ground truth, large areas can be accurately surveyed for a particular purpose. The initial task of survey and appreciation of the landscape is accomplished by using the remotely sensed data obtained from the Indian Remote Sensing satellites. It is axiomatic that in assessing the potential use of land, the existing land use must be taken into account in order that the feasibility of improving production can be assessed.

The approach of integrated surveys was the outcome of the concept of division of land into units of natural complexes, which are repeated, in regular patterns in the landscape. However, to delineate such units, information on the physical and social aspects of the land are to be considered with all their complexity. The approach to integration is the compilation and collection of information about several resources factors obtained from different sources.

The important task of preparing an action plan map by integrating different mono-disciplinary information was done using geographic information system. As said earlier in the process of planning, many modifications may be needed as the plan is being implemented. The availability of resource information in the computer for a long time makes it easier for planners to access the information in its previous form and also for obtaining outputs at will. GIS has become an invaluable tool in the planning set-up.

Integration of information being the sole operation involved in this study, an action plan can formulated for development of land resources and eventually for land use planning.

Acknowledgement

The paper is based on project work completed by Mr Rahul Tiwari under the supervision of Prof. P.R. Vyas, Director NRM, JRN RVU, Udaipur (Internal Guide) and Dr V. Madhava Rao, Professor and Head, NIRD, Hyderabad.

References

Anuta, P.E. (1977), 'Computer Assisted Analysis Techniques for Remote Sensing Data Interpretation', *Geophysics*, Vol. 42, pp. 468–81.

Campbell, J.B. (1983), *Mapping the Land Aerial Imagery for Land Use Information*, Association of American Geographers, Washington, DC.

Jensen, J.R. (1983), 'Urban-Suburban Land Use Analysis', Chapter 30 in R. N. Colwell (ed.), *The Manual of Remote Sensing*, Vol. 2, American Society of Photogrammetry and Remote Sensing, Falls Church, pp. 1571–966.

Lindgren, David T. (1985), *Land Use Planning and Remote Sensing*, Martinus Nijhoff Publishers, Boston, pp. 101–15.

Rao, D.P. (1995), *Remote Sensing for Earth Resources*, Association of Exploration Geophysicists, Osmania University, Hyderabad, pp. 153–69.

Vink, A.P.A. (1975), *Land Use in Advancing Agriculture*, Springer-Verlag, Berlin, p. 394.

13

Application of Remote Sensing and GIS Technique for Land Use Planning in Irrigation Project Area

A Case Study of Upper Betwa Basin of Madhya Pradesh

P.R. Vyas and Shankar Bagale

Irrigation systems often have a number of competing objectives and are assessed by interest group with differing values and perspectives; a wide range of performance indicators is thus required. Performance indicators do more than measure the value of a particular item such as yield or canal discharge. They have to include a measure of quality as well as of quantity and be accompanied by appropriate standards or permissible tolerances; if the value of the indicator falls outside a particular range of values then performance is presumed to be unsatisfactory. The importance of particular indicators depends on the relative scarcity of land and water, and the cropping patterns and sequences. The ultimate utility of a particular set of performance parameters is that they guide managers into better performance in the future, because they facilitate judgment as to the level of performance actually achieved and the underlying causes of that level of performance.

The analysis of the irrigated crop inventory data together with the basic field data on Culturable Command Area information, water drawls/releases into canals facilitating estimation of irrigation system performance indicators are described in this chapter.

Cropping Pattern

Cropping pattern in the command area determines the irrigation needs and hence the analysis of cropping pattern data is useful to know whether the irrigation system is optimally used or not. Every irrigation system has a designed cropping pattern and accordingly water deliveries are let into the system. But in practice, generally it is observed that the farmers do not adhere to the designed cropping pattern and hence has direct bearing on the water management in the command areas. Satellite derived total crop area, cropping pattern (paddy and others) in both the study years was analysed to identify the changes in the cropping pattern, reduction in crop areas, and possible reasons for the changes.

Irrigation Intensity

The intensity of irrigation is defined as the percentage of crop area and the culturable command area. Usually the area to be irrigated during each crop season is expressed as a percentage of the culturable commanded area, which represents the intensity of irrigation for that crop season. It indicates the gap between the irrigation potential created and potential utilized. By knowing the intensity of irrigation for a crop season, the area to be irrigated during that crop season can be determined. Further, by knowing the areas over which different crops are grown during any crop season as well as the supply of water for each of these crops, the water requirement for the crop season can be determined. Since this signifies the acreage that can be irrigated from the available water, it largely governs the design and operation of irrigation system and hence the water management in the command area. Irrigation intensity was estimated at Krishna Western Delta (KWD) level and disaggregated level using the crop area extraction methods using the irrigation command boundary as base.

Uniformity of Water Application

The uniformity indicator refers to the variation (or non-uniformity) in the amounts of water applied to locations within the irrigated area. It is defined as below:

$$Uniformity\ Index(-) = \frac{V_i / A_i}{V_1 / A_1}$$

where:

V_i = total irrigation volume supplied to KWD (m^3);

V_i = irrigation volume supplied to command i (m^3);

A_i = total irrigated area KWD (ha);

A_i = irrigated area in command i (ha).

Water Productivity

The water productivity indicator quantifies the yield per volume of irrigation water supplied. Comparing this indicator for different command areas, it illustrates the spatial variation in productivity. With this, the relation between salinity and productivity can be studied. The water productivity is defined as:

$$Water\ productivity\,(kg\,/\,m^3) = \frac{V_i}{V_1}$$

where

Y_i = Paddy yield in command i (kg);

V_i = Volume of irrigation water supplied to command i (m^3).

Water Utilization Index and Depth of Water

Water Utilization Index (WUI) is defined as the ratio of the crop area irrigated in an irrigation command and the water delivered to that irrigation command. In a broader sense, this is a measure of effective utilization of water in an irrigation command and this index enables to compare the performance of irrigation system across the command, i.e., at disaggregated level. This is also a measure of how extensive or intensive the irrigation is during any crop season or annually.

Data on monthly water drawls into the canals that offtake from the main canal were obtained for crop years 1998–99 and 2005–06 This was used to arrive at the monthly water drawls at the head of irrigation canal to calculate the seasonal irrigation supplies and aggregated to crop growth period (Kharif).

Estimation of WUI involves converting the areas under various crops in terms of a single major crop grown in the command area. In this present study, paddy and annual crops were brought at par with wet crop (paddy) by taking operationally used equivalent wet crop conversion factors given below. The conversion factors were estimated based on the supply of water adopted from time to time in the project area taking into consideration of crop water requirements and irrigation efficiencies in the irrigation system.

Equivalent Wet Area: Area of (Paddy + 1.5*Annual crops)

WUI values were estimated using equivalent wet area and canal discharges for each irrigation command. Delta (the reciprocal of WUI) was estimated for each of the irrigation blocks. WUI and delta, i.e., depth of irrigation water supplied to each irrigation block level were used in analysing the spatial irrigation service performance across the command area during the selected time periods.

Methodology

In the present study two objectives are taken up:

(i) Base line inventory using Remote Sensing and Geographic Information System technology for feasibility assessment of new irrigation project.

(ii) Irrigation Performance Assessment of KWD command area.

These are discussed separately as follows:

Objectives

The study considered for base line inventory for feasibility assessment of new irrigation project on Betwa River at Upper Betwa Basin of Madhya Pradesh, India involves:

- Land Use and Land cover mapping.
- Submergence area analysis of the proposed dam at Barrari.

Ken-Betwa Link (K-B Link)

Ken-Betwa link project is the diversion of surplus water of Ken basin to water deficit Betwa basin. This surplus water is utilized for irrigation, drinking and other domestic use in Madhya Pradesh and Uttar Pradesh. K-B link project enroutes the irrigation of 47,000 ha in Chhatarpur and Tikamgarh districts of Madhya Pradesh and Hamirpur and Jhansi districts of Uttar Pradesh of Upper Betwa Basin with 1,375 m^3 of water planned to be diverted. According to prepared indicative master plan of Betwa basin, there are four irrigation projects namely Barari, Neemkheda, Richhan and Kesari dams, which lie in the Upper Betwa basin. Annual irrigated area under these four projects is 1.27 lakh ha and the water use is 659 m^3. Therefore, 659 m^3 of water is proposed to be diverted from Ken to Betwa river at upstream of Parichha weir through the K-B link. The water diverted to the upstream of Parichha weir will meet the water requirement of its downstream commands. The water thus saved is proposed to be utilized in the upper reaches of Betwa through these four projects by way of substitution. The command area under these four projects has been termed as "Betwa Command".

Barrai and Kesari dam are falling in the study area in which only Barrari dam is considered for submergence analysis.

Betwa River

The Ken and Betwa basin are the part of Ganga Basin in Madhya Pradesh. The Betwa river originates in the Raisen district of Madhya Pradesh near Barkhera village south-west of Bhopal at an elevation of about 576 metres above mean sea level. It flows in a north-easterly direction through Madhya Pradesh. The total length of the river from its origin to confluence with the Yamuna is 590 km. The river joins the Yamuna near Hamirpur in Uttar Pradesh at an elevation of about 106 m. The river basin lies between the latitudes of 22°54′ N to 26°00′ N and the longitudes of 77°10′ E to 80°20′ E. The total catchments area of the basin is 43,895 sq km. It covers the districts Tikamgarh, Sagar, Vidisha, Raisen, Bhopal, Guna, Shivpuri and Chhatarpur of Madhya Pradesh and Hamirpur, Jalaun, Jhansi, Banda districts of Uttar Pradesh. During its course from the source up to the confluence with the Yamuna, the river is joined by a number of sub-tributaries, the important among them being Bina, Jamini, Dhasan and Birma on the right bank and Kaliasote, Halali, Bah, Saga, Narain and Kaithan on the left bank.

Geographic Location

The study area covers the Upper Betwa Basin area confined between 77° 45′ 00″ to 78° 30′ 00″ East longitude and 24° 30′ 00″ to 23° 30′ 00″ North latitude. The extent of study area (covering approximately 4.17 lakhs ha) and topo-coverage is given in Plate 13.1.

Out of proposed 4 dam sites, location of 2 dam sites falling in the study area are at Barari: 23°40′30″E 77°50′30″N Kesari: 23°52′32″E 78°01′34″N.

Satellite Coverage

The Upper Betwa Basin command considered for the study falls in number of LISS IV of IRS P6. The satellite coverage is as shown in Plate 13.1.

Existing Land Use Pattern

The study area comes under Vidisha, and Raisen districts of Madhya Pradesh that covers 90 per cent area of the region. The general land use pattern of these districts is as given in Table 13.1.

Table 13.1

General Land Use Pattern of Vidisha and Raisen Districts

In ha	*Raisen*			*Vidisha*		
	1978–79	*1988–89*	*1998–99*	*1978–79*	*1988–89*	*1998–99*
Geographical Area	**848746**			**730197**		
Net Area sown	402800	418400	426247	502000	517300	530639
% Net to Geographical area	47.5	49.3	50.3	68.7	70.8	72.7
Forest	848000	333800	333407	730000	105100	105825
Land not available for cultivation	38000	41500	42212	47000	45700	46569
Uncultivated land excluding fallow land	27000	28000	29062	43000	39500	31964
Cultivable waste-land	38000	21100	13842	25000	15200	10259
Fallow land	8000	5900	3973	8000	7400	4941
Total Waste land		834.55			1132.61	

Observed cropping pattern in Upper Betwa Basin of Madhya Pradesh for all four proposed project command is as given in Table 13.2.

Table 13.2

Cropping Pattern in Upper Betwa Basin

S. No.	*Crop Type*	*% Cropwise*
1	Paddy	1.01
2	Jawar	8.83
3	Wheat	48.52
4	Maize	1.58
5	Pulses	26.98
6	Sugarcane	0.2
7	Vegetables and Fruits	0.29
8	Other Food Crops	1.2
9	Oilseeds	4.04
10	Fodder	7.35

Methodology

Methodology adopted for base line inventory for irrigation project is as given below:

Plate 13.1

Georectified IRS P6 LISS IV Images of Upper Betwa Basin Acquired During 2005-06

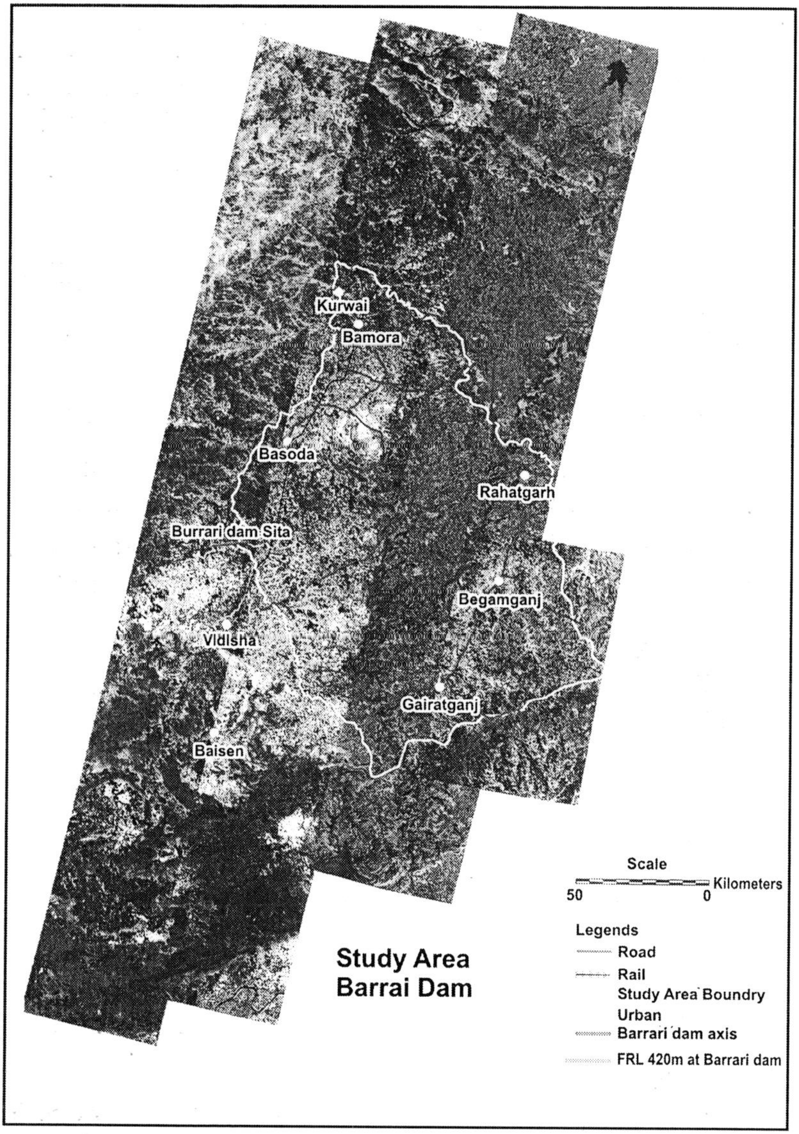

- Satellite Data selection and rectification
- Generation of base map
- Image interpretation and digitization
- Land use and land cover (LULC) patterns
- Area of submergence for dam
- Land use and land cover statistics

The study involves pattern reorganization from georectified satellite data of the study area. These recognized patterns, digitized in each category, such as built-up lands, agriculture, forest, wastelands, water bodies, etc., are taken into GIS layer for generating LULC statistics. And also fixing up Forest Research Laboratory (FRL) of dam on this LULC map and submerged area statistics are calculated.

Satellite Data Selection

The study area covers approximately 4.18 lakh ha of Upper Betwa Basin. Satellite selection is decided on the basis of:

- Level of LULC classification
- Date of pass, i.e., Temporal resolution
- Availability of data from India and other countries' satellites

In this study 2nd level of LULC classification was considered. For interpreting these 2nd level LULC classification features from an image requires some range of spatial resolution. Therefore, images from Indian satellite IRS P6, LISS IV is suitable which has 5.8 m spatial resolution within 24 days receptivity. The available cloud-free satellite data are browsed for 2005–06 season for the considered study.

The study area covers 33 scenes of LISS-IV of IRS-P6. Table 13.3 provides data acquired through satellite on particular path, row through images.

Table 13.3

Details of Satellite Data Acquired

S. No.	*Strip No.*	*No. of Images*	*Satellite*	*Path No.*	*Row No.*	*Date of Pass*
2	5	8	IRS-P6	201	88 to 95	20.04.06
3	6	8	IRS-P6	102	64 to 71	06.04.05
4	7	7	IRS-P6	202	68 to 74	14.01.06
5	9	1	IRS-P6	102	68, 69	12.04.05
Total	33					

Rectification of Satellite Data

The selected LISS-IV data of IRS-P6 for 2005–06 covering the study area was geo-referenced to base map using ERDAS software, by identifying ground control points. Steps followed for rectification are:

- Monodate LISS-IV image were registered with each other in a single strip. A single strip image is obtained by mosaiking all these monodate images. From 33 images 5 strips are generated.
- Selecting the central strip as master strip and registered each strip with this central strip. And single mosaic is obtained from these 5 strips.
- This single LISS-IV mosaic is then georectified with ETM Pan merged data of 14.5m spatial resolution taking 63 GCP points and resampled (spatial interpolation) to 5.8 m resolution using 3rd order polynomial transformation function because the RMS error is exceeded from sub-pixel level. But to maintain spatial characteristics and orientation of features between these multidate LISS-IV strips.
- For resampling nearest neighbourhood method is used. The generated digital output was used as a reference image for geo-referencing other dates of satellite data. The considered datum is UTM 43 zone of North spheroid WGS 1984. The same rectification steps are adopted for rectifying scanned ancillary maps.

References

Aronoff, S. (1985), *Geographic Information Systems: A Management Perspective,* WDL Publications, Ottawa, Canada.

Bose, M.G. (1991), 'Performance Evaluation of CAD Projects', *Irrigation and Power Journal,* p. 61.

NRSA (2005), *Manual of Procedure for Wastelands Using Remote Sensing Technique,* National Remote Sensing Agency, Hyderabad.

—— (2005a), *Satellite Remote Sensing Based Evaluation Study of Nagarjunsagar Left Canal Irrigation Command, Andhra Pradesh State,* National Remote Sensing Agency, Hyderabad.

Sabins, F.F. (1987), *Remote Sensing: Principles and Interpretation,* 2nd Edition, W.H. Freeman & Co., New York.

14

Land Use/Land Cover Analysis of Teetriya Gram Panchayat Using GIS and Remote Sensing Techniques

P.R. Vyas, Ashwini Kumar Vaishnav and J.P. Sharma

Land resources are considered as a base for various developmental activities on the earth. As we have observed there is an ever-increasing pressure of population on land for meeting the growing demand for food nowadays. This is applicable particularly in areas of erstwhile barren, fallow, marginal and forest land.

Land use/land cover assessment is one of the most important parameters to meaningfully plan for land resources management. Land use/land cover inventories are assuming increasing importance in various resource sectors like agricultural perspective study. Land use/land cover changes permit identification of long-term trends in anticipation of the problems that accompany change in land use. Information of land use/land cover also provides a better understanding of the wasteland and surface water bodies, which is vital for development and planning. GIS and remote sensing, the two mutually compatible technologies can be used more effectively in natural resources management. The timely, accurate and up-to-date information on land use/land cover can be obtained from various Remote Sensing satellites based on high-resolution, remotely sensed data on a cost effective basis, at the shortest possible time.

Scope of the Study

Information on land use/land cover is the basic prerequisite for land resources evaluation, its utilization and management. A considerable degree of land transformations is being witnessed as a result of growing population pressure on the finite land resources, consummating in deterioration of the environment. As a precursor, it is necessary to understand the cause and effect of the transformations through scientific studies.

Status of socio-economic and agricultural growth of a country depends on the availability of surface and groundwater. In India, more than 80 per cent of rural and nearly 20 per cent of urban population mostly depends on groundwater for meeting their drinking and domestic requirements. The distribution of groundwater is not uniform in all the regions. The spatio-temporal variations in rainfall and regional/local differences in geology and geomorphology have led to an uneven distribution of groundwater in different regions across the country. This uneven distribution (poor prospects) and indiscriminate tapping and also over-exploitation in certain zones are the main reasons leading to scarcity of groundwater in many parts of the country.

The scope of the present study is to prepare the thematic layers of land use/land cover pattern, their assessment, spatial distribution and extent, using remote sensing and GIS techniques for Teetriya Gram Panchayat, Chakshu Block, Jaipur District. Considering the requirements of water for drinking and irrigation, a detailed investigation for both surface and groundwater has been carried out using satellite and ground-based information.

Objectives

Following objectives have been taken into account in the study:

(i) To analyse and prepare the following themes on 1:12,500 scale using remote sensing and GIS techniques.

(ii) To create base map (settlements, village boundary, road, etc.) layer.

(iii) To make land use/land cover layer through GIS techniques for its analysis.

Study Area

Teetriya Gram Panchayat of Chakshu Block is located in Jaipur District of Rajasthan and comprises of seven villages, viz.; Teetriya, Sri

Krishnapura, Brijnathpura, Bhojyand, Chak Srikishanpura, Deo Kishanpura and Sri Daulatpura. The area of study is bounded by 26°37′26.66″N to 26°41′1.69″N latitude and 75°48′53.53″E to 75°53′8.97″E longitude. The area of study lies on the west of Jaipur-Tonk National Highway, 28 km south-south-east of Jaipur city.

In general, the topography of the area is flat with slight undulating at some places. Virtually, the area is devoid of any prominent drainage except the development of a few ravine type of drainages in the eastern part (just outside the study area). However, there are numerous surface water bodies (filled during rainy season) of varying dimensions in different villages. A large water reservoir is present near Chandlai village (outside the study area) in the north-east of Teetriya village. The major part of the study area is under agricultural use. Oran/Gochars with Nadi (small surface water bodies) exist in many places in the study area.

The population and geographical area of villages under Teetriya Gram Panchayat as per 2001 Census is in shown in Table 14.1.

Table 14.1

Population (2001) and Area in Teetriya Gram Panchayat Project Area

S. No.	*Village*	*Population*	*Area (Sq km)*
1	Bhojyand	494	1.78
2	Brijnathpura	618	1.96
3	Chak Shri Kishanpura	34	0.50
4	Dev Kishanpura	948	4.16
5	Shri Doulatpura	24	0.845
6	Shri Kishanpura	361	1.03
7	Teetriya	2,908	12.28
	Total	5,329	22.56

Temperature and Rainfall

It has been observed that the minimum and maximum temperature varied from 3°C to 45°C in the last ten years (1995–2005) and the annual rainfall varied between 45 mm and 681 mm. The average annual rainfall of the area is 548.2 mm. The monthly rainfall data (in mm) for 2003, 2004 and 2005 is shown in Table 14.2.

Table 14.2

Monthly Rainfall Data (in mm) for Teetriya Gram Panchayat Area

Month	*2003*	*2004*	*2005*
January	10.8	2.5	0.7
February	45.1	0	0
March	0	0	3.4
April	0	0.7	24
May	0	32.3	–
June	94	23.3	–
July	148	146.5	–
August	149	413.2	–
September	67.8	27.7	–
October	0.1	33.7	–
November	0	0.4	–
December	13.1	0	–
Annual RF (mm)	527.9	680.3	28.1

Methodology and Approach

The methodology attempted here is based on visual and hybrid (digital combined with visual) classification of RESOURCESAT (IRS-P6) LISS-IV standard false colour composite (FCC) generated using band 2, 3 and 4 respectively and also image characteristics like tone/colour, texture, pattern, shape, size, location, shadow and association (see Figure 14.1).

Base Map Preparation

The base map is referred to as a map containing the basic reference details of infrastructure details, viz.; road and rail, which act as points of reference while transferring the thematic details from topographical map on 1:50,000 scale. The base details are further updated using satellite image.

Development of Interpretation Key

It essentially consists of a set of image elements or characteristics, which help in the recognition or identification of various land use/land cover classes systematically on the satellite imagery for visual interpretation and digital classification. The preliminary interpretation key prepared for satellite data has been finalized after the ground truth, whenever it was found necessary. However, the interpretation key may differ with the change in type of satellite data, band combination, season, scale, etc.

Figure 14.1

Methodology to Measure Land Use/Land Cover

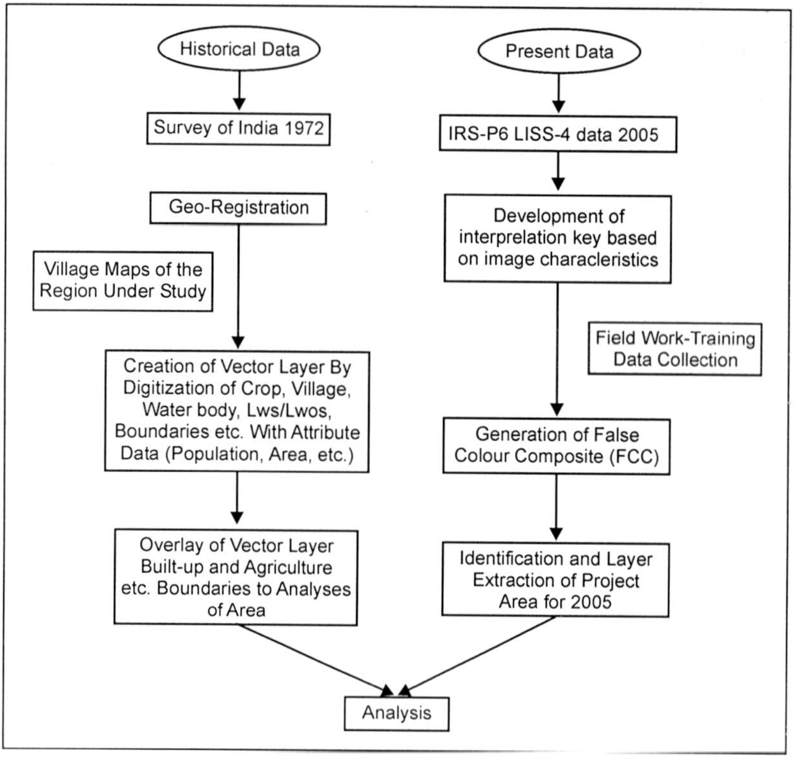

Final Map Preparation and Area Estimations

The land use/land cover map is prepared from the digitally enhanced and geo-referenced satellite imagery (IRS-P6) using hybrid classification technique. Digital classification has been performed based on classification system mentioned in the Table 14.3 and validated with the preliminary pre-field interpretation and ground truth data collection. The Integrated Land use/cover map of the study area is shown in the Plate 14.1.

The land use statistics is generated for the study area is shown in the Table 14.3; and its graphical representation is shown in Figure 14.2. The Table 14.4 shows crops and settlements in the study area.

Table 14.3

Land Use/Land Cover Estimates for Study Area

S. No.	*Land Use/Land Cover Level-1*	*Land Use/Land Cover Level-2*	*Land Use/Land Cover Level-3*	*Area (in sq km)*	*Percentage Area*
1	Built-up Land	Built-up Land	Rural	0.38	1.68
2	Agricultural Land	Crop Land	Kharif Crop	6.53	28.95
			Rabi Crop	3.52	15.60
			Double (Kh+Rb) Crop	5.76	25.53
		Fallow Land	Current Fallow	3.39	15.03
3	Wasteland	Land with Scrub	Land without Scrub	1.83	8.11
4	Water Bodies	Tanks/Pond	Tanks/Pond	0.21	0.93
		Total		22.56	100.00

Table 14.4

Crops and Settlements in the Study Area

S. No.	*Category*		*Description*
1	Crop	Kharif Crop	*Bajara, Mung, Til, Gwar, Tinda, Udad, Chola,* Groundnut, Sunflower
		Rabi Crop	Wheat, Jow, Gram, Mustard, Tara-meera, Rajka
2	Settlement		Teetriya, Shri Daulatpura, Deo Kishanpura, Bhojyand, Shri Kishanpura, Brijnathpura, Chak Shri Kishanpura

Observations and Assessment of Land Use/Land Cover

The Teetriya Gram Panchayat comprises an area of 22.50 sq km. The area under various land use/land cover categories within study area are computed in Table 14.3.

The major land cover of the study area is under kharif crop land, i.e., 6.53 sq km (28.95%) followed by double crop, i.e., 5.76 sq km (25.53%) of the total study area. In the present report the figures/numbers of `area' and `percentage' are calculated from the total study area, i.e., 22.56 sq km.

The land use/land cover categories such as built-up land, agriculture, water bodies, wasteland and other have been identified. The study area is under various land use. Integration of remote sensing and GIS can give a better solution for natural resources management problems. Satellite imagery is more convenient and economical than other options of getting

Figure 14.2

Land Use/Land Cover of Teetriya Gram Panchayat

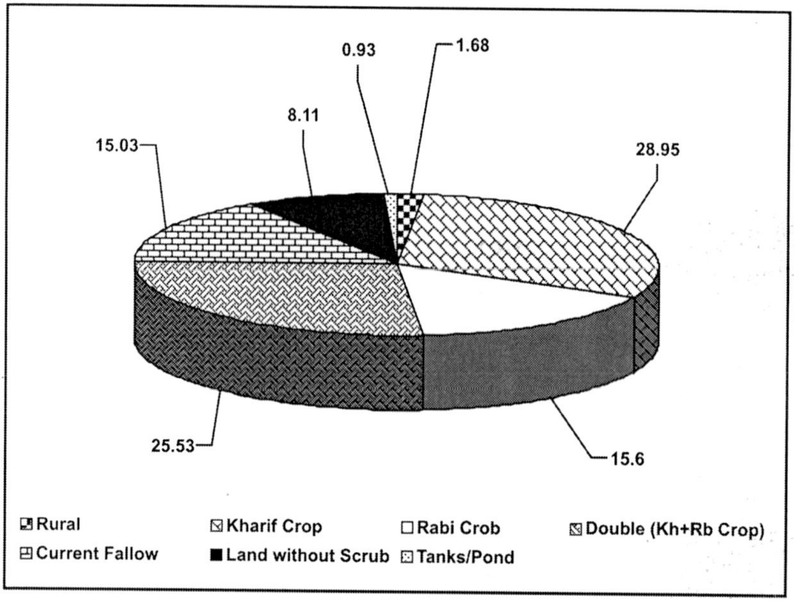

information. Analysis of digital data provides us more accurate and effective information, on various natural resources. People utilize the land for various agricultural practices. Underutilization, overexploitation and mismanagement could be observed in the field. As a consequence, the yield from agriculture is not optimum.

Integration of remote sensing and GIS can give a better solution for natural resources management problems. Satellite imagery is more convenient and economical than other options of getting information. Analysis of digital data provides us more accurate and effective information, on various natural resources.

In the present study the land use/land cover mappings has been done using visual image interpretation classification technique. The results of the study are discussed in the following paragraphs.

Built-Up Land

It is defined as an area of human habitation developed due to intensive non-agricultural use. Village is also the only class under the built-up land category, that can be included in total number of seven built-up land in the study area. Four villages could be clearly

Plate 14.1

Teetriya Gram Panchayat and Surrounding Area

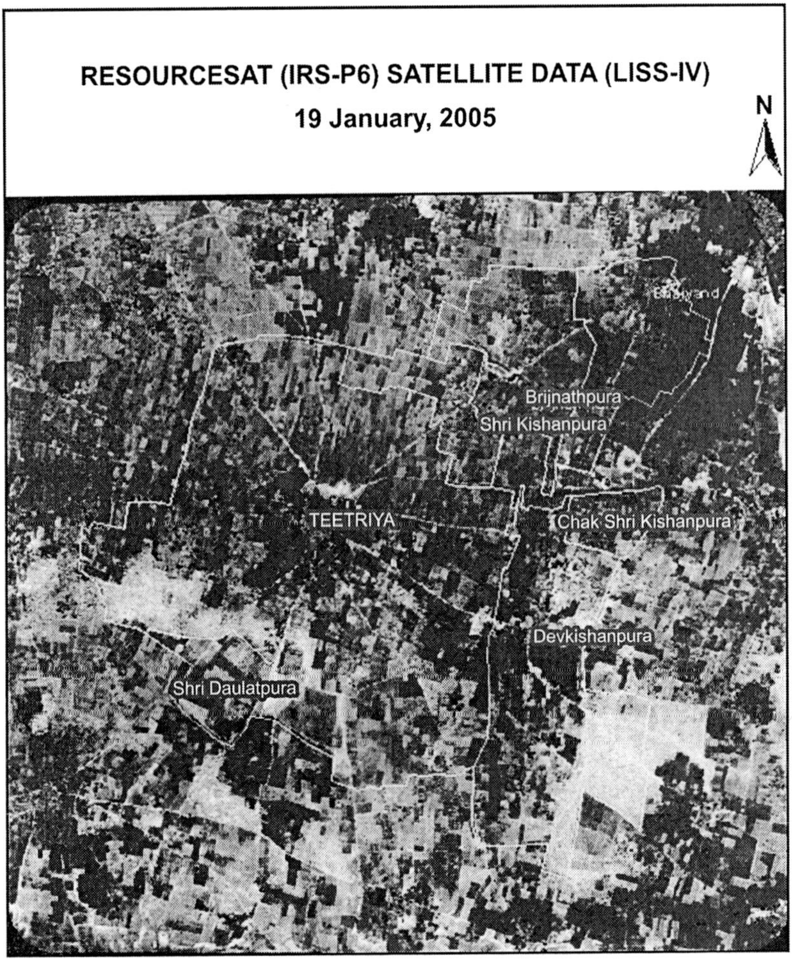

identified from the satellite data and two villages (Shri Kishanpura & Shri Doulatpura) are not identified. The village Teetriya comprises of few houses, interspersed with trees and agricultural field. The major category discernible on the satellite data in the study area including residential, roads, etc., are identified and mapped. The residential areas (settlements), confined within the study area estimated are 0.38 sq km (1.68%) (Table 14.3).

Plate 14.2

Teetriya Gram Panchayat and Surrounding Area

Agricultural Land

From the satellite data the agriculture area could be clearly delineated into two categories, such as single crop and double crop. Crops raised in the land once in a year (Kharif and Rabi) are considered to be single crops. The area occupied by this class during 2005 is shown in Plate 14.1. Cropland is the land under cultivation. In the image bright red tone, smooth texture and square shape easily identifies the cropland. In the present study the cropland category includes standing/harvested cropland in Kharif season. The percentage of

Kharif use is highest in part of Teetriya, and Rabi use is highest in part of Teetriya and Bhojyand. The total cropland in the study area is estimated to be Kharif 6.53 sq km (28.95%), Rabi 3.52 sq km (15.60%), double crop 5.76 sq km (25.53%) as the satellite data used in the study pertained to January (Rabi) – August (Kharif) month. The rest of cropped area where the crop is already harvested is categorized in the fallow category; the major crops grown in Kharif season are, *jowar, bajra, til, moong, urd, arhar*, and in the Rabi season are wheat, *sarso, muttor, chana, masur*, etc. Paddy is normally taken as a mono crop while others are mixed sown (Table 14.3).

Fallow Land

Fallow land is the agricultural land in which the crops are taken out and kept uncropped for one or more seasons but not less than one year. In the image, the fallow land was easily identified by greenish gray to greenish blue tone, smooth texture and square shape. The percentage of fallow land use is highest in part of Teetriya. The total fallow land in the study area is estimated to be 3.39 sq km (15.03%) of the study area (Table 14.3).

Wastelands

Among the wasteland category there are only two classes namely, land with or without scrub. Wasteland is one of the important categories which is used for other than agriculture purposes, in the study area. Wasteland is described as degraded land which can be brought under vegetative cover with reasonable effort and which is currently underutilized for lack of appropriate water and soil management or on account of natural causes (NWDB, 1991). The wastelands type identified in the study area is land with/without scrub. These categories are classified and shown in Table 14.3. The total area under wasteland category estimated is 2.77 sq km (12.28%) of the total study area.

Land with Scrub and Land without Scrub

This category appears on satellite imagery as light yellow to brown to greenish blue patches of varying sizes with irregular shapes and are usually associated with uplands. Hence, area with scrubs has been separated from those that are devoid of any such vegetation cover, which enables us to classify them separately. The percentage of land with scrub is highest in part of Teetriya and Shri Doulatpura and land without scrub is highest in part of Dev Kishanpura. In the study area, land with scrub area estimated is 0.94 sq km (17%) of the total area and land without scrub area estimated is 1.83 sq km (8.11%) of the total area (Table 14.3).

Waterbodies

Ponds and tanks are the important water bodies seen in the study area. The area statistics for this category is 0.21 sq km (0.93 per cent). The spatial extent under waterbody is depicted in Table 14.3.

Conclusions

The land use/cover assessment using satellite imagery provides reliable and accurate information, which is cost- and time-effective. It also offers a holistic view of large areas for better monitoring of land use/cover occurrence and distribution. Hence, satellite remote sensing and GIS techniques are useful tools for assessing the land use/land cover. The satellite imagery due to its repetitive advantage is also useful for undertaking monitoring of changes in land use patterns.

Acknowledgement

The paper is based on project work carried out by Mr Ashwini Kumar Vaishnav that was supervised by Dr Rakesh Paliwal, Scientist at RRSSC, Jodhpur and Dr P.R. Vyas (Internal Guide), Director, NRM, JRN Rajasthan Vidyapeeth, Udaipur.

References

Anderson, J.R. (1971), 'Land Use/Land Cover Changes: A Framework for Monitoring', *Journal of Research*, Vol. 5, No. 3, U.S. Geological Survey, Government of India, pp. 143–53.

Anuta, P.E. (1977), 'Computer Assisted Analysis Techniques for Remote Sensing Data Interpretation', *Geophysics*, Vol. 42, pp. 468–81.

Aronoff, S. (1982), 'Classification Accuracy: A User Approach', *Photogrammetric Engineering and Remote Sensing*, Vol. 48.

Bisht, B.S. and B.P. Kothyari (2001), 'Land-Cover Change Analysis of Garur Ganga Watershed Using GIS/Remote Sensing Technique', *J. Indian Soc. Remote Sensing*, Vol. 29, No. 3, pp. 165–74.

Campbell, J.B. (1983), *Mapping the Land Aerial Imagery for Land Use Information*, Association of American Geographers, Washington, DC.

Hord, R. Michel (1982), *Digital Image Processing of Remote Sensing Data*, Academic Press, New York.

Jensen, J.R. (1983), 'Urban-Suburban Land Use Analysis', Chapter 30 in R. N. Colwell (ed.), *The Manual of Remote Sensing*, Vol. 2, American Society of Photogrammetry and Remote Sensing, Falls Church, pp. 1571–966.

Lecture Notes, IIRS, Dehradun.

Lecture Notes, RRSSC, Jodhpur.

Lillesand, T.M. and R.W. Keifer (1979), *Remote Sensing and Image Interpretation*, John Wiley & Sons, New York.

Lo, C.P. and Albert K.W. Yeung (2002), *Concepts and Techniques of Geographic Information System*, Prentice Hall of India Pvt. Ltd., New Delhi.

Vink, A.P.A. (1975), *Land Use in Advancing Agriculture*, Springer-Verlag, Berlin, p. 394.

15

Application of Remote Sensing and GIS Techniques for Land Use and Land Cover Analysis and Change Detection Study of Udaipur City

P.R. Vyas and Deepak Kumar Ahuja

Importance of land as economic resource, whether rural or urban, is long recognized. Efforts are therefore directed to achieve the best possible use of land under a given situation. Several tools and techniques have been developed from time to time to bring about efficient utilization of land. Mapping organization of database and assigning the use or zoning are major elements of land management. Studying temporal changes in the use of land, assessing the future needs and promoting the land use accordingly are the core activities of land-use planning.

With the rapid pace of urbanization combined with the explosive population growth, urban areas and its surrounding regions are undergoing a dynamic change. Because of this process, there is an increasing demand on urban land and services. As the limited land resources in the city gets used, the pressure mounts on the surrounding environs of fertile, unsuitable and vulnerable lands around the city, causing faster rate of land conversion from non-urban to urban use.

The natural resources are responsive to human interaction and these together with tertian feature determine the selection of proper land-use pattern which reflects the cultural, social and economic conditions (Vink, 1975). Most of the natural resources are directly or indirectly related to the surface cover in a given locality. The changes in land use are largely related with the external forces and pressure built-up within the system (Bisht and Kothyari, 2001).

Advances in computer technology have contributed to the technique of land-use planning in numerous ways. The use of remote sensing data and image interpretation techniques coupled with geographic information system (GIS) tools is widely used for mapping of existing land use as well as constant monitoring and subsequent preparation of urban plans. With respect to urban areas, remote sensing data products providing almost real time information are not only valuable tools but also cost- and time-effective. An immense scope of remote sensing and GIS techniques in monitoring urban expansion, land use mapping change detection in various cities of world have been shown by various workers.

The launch of Indian IRS-1C & 1D satellite and PAN sensors have provided an opportunity to derive the land use/land cover information at regional, local and parcel levels on appropriate scales with an assumed frequency for updating. Raghavswamy, et al., (1996) have demonstrated that by merging IRS-1C, LISS III and PAN data, various details of urban land use classes of level III can be identified which is advancement over IRS-1A/1 B, LISS II data of 36 metres resolution.

Objectives of the Study

The aim of the present project work is land use/land cover mapping using satellite remote sensing digital techniques and GIS supplemented with field survey to carry out change detection studies in the selected area.

ArcView software is used to create derived maps from the primary thematic maps stored as vector form in the database. ERDAS software is used for the supervised classifications.

The study has been carried out with the following objectives:

- To generate the base map for the study.

- To prepare the map of land use/land cover classification for the study area using the satellite remote sensing data and supervised classification algorithm.
- To assess the change in land use/land cover in the study area.

Study Area

The study area is selected from the southern part of the Udaipur district. It lies between 24°30′N to 24°35′N latitudes and 73°40′E to 73°45′E longitudes. The total geographical area of the district is 14.62 lakh ha (27% of the state). It lies at an elevation of 560 m above MSL generally increasing towards west. It abounds in hills, some are steep in the south-west corner and wooded in north-west and south-east. The northern portion is characterized by elevated plateau. The eastern portion is fertile plains. Main rivers of the region are the Banas, Berach, Som, Jakham, Sabarmati and Wakal.

Physical Background of the Study Area

The study area comprises of lakes, settlements and forest/vegetation. The district is flanked by Dungarpur, Banswara, Chittorgarh, Rajsamand and Sirohi district of Rajasthan and Banaskantha district of Gujarat. It is administratively divided into ten tehsils. The district forms a part of Aravali landscape. It is mainly discontinuous hilly tracts in the south (88%) and upland (12%) in the eastern part. The geological formations are Aravali metamorphics comprising micaschist, quartzite, dolomite, hornblende, schist, gneisses, etc. Vegetation of the district belongs to tropical dry deciduous type.

The climate of the district is tropical semi-arid, characterized by dry hot summer and intense weather. The average rainfall is 650 mm and of the total rainfall about 90–95 per cent occurs during June to September, the maximum being in July. Variation of annual rainfall is observed from year to year.

The winter sets from December and continues to February. The hot season commences form March and ends with the middle of June. The monsoon season is spread over mid-June to mid-September. The minimum and maximum temperature has varied from 3°C to 47°C in the last ten years.

Methodology and Approach

Image classification is a procedure to automatically categorize all pixels in an image of a terrain into land cover classes. Normally, multi-spectral data are used to perform the classification of the spectral pattern present within the data, for each pixel is used as the numerical basis for categorization.

The digital processing of satellite images are used for land use/land cover analysis and change detection studies. The satellite images of IRS (LISS-III) and Landsat (TM) data are used in the present study for land use/land cover analysis of Udaipur city (south). The various thematic layers are prepared using the Survey of India toposheet. The whole study was carried out in following stages:

(i) Data collection and interpretation/mapping.

(ii) Data verification.

(iii) Map finalization/Database creation.

(iv) Data Analysis.

Figure 15.1

Methodology to Measure Land Use/Land Cover

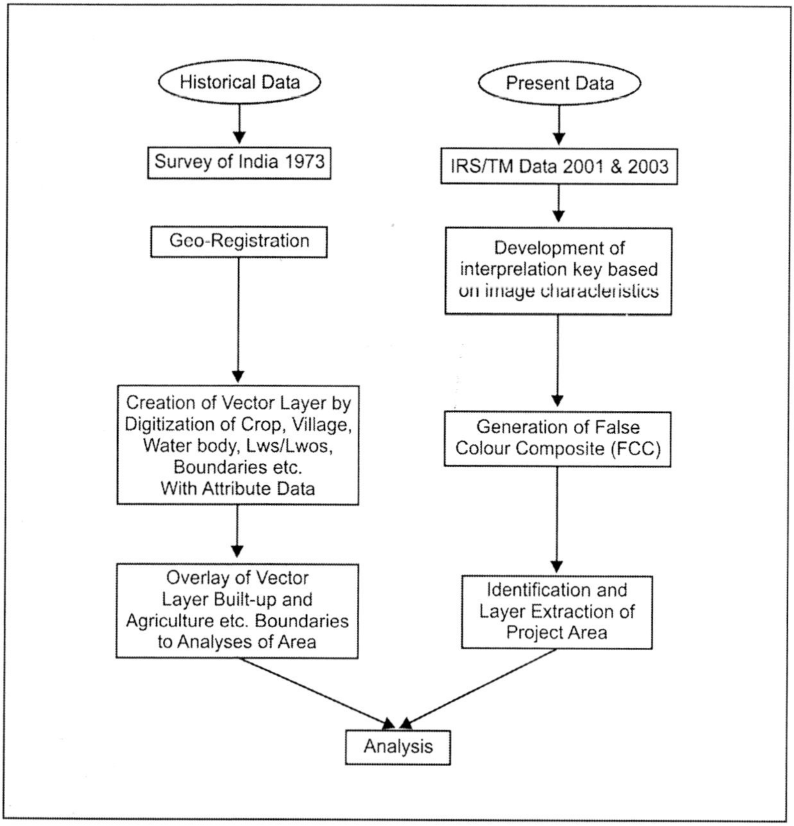

Selection of Parameters for Land Use/Land Cover

To develop any land use/land cover classification system, it is essential to consider certain criteria in order to arrange it under a suitable framework to facilitate systematic land use inventory, mapping and change detection studies. The important features used for analysis are: agriculture land; natural Vegetation; settlements; bare land; dense scrub land and water body.

Source of Satellite Data

Major sources of satellite data are as under:

- Satellite data of Landsat (TM), January 2001.
- Satellite data of IRS (LISS-III), January 2003.
- Survey of India (SOI) Topographical map (45H, 10) on 1:50, 000 scale for preparing different thematic layers.
- Ground truth data for study areas using GPS.

Geo-referencing of Satellite Data

The satellite data of both IRS and Landsat are geo-referenced to bring it to the same coordinate system. Both imageries are in Lambert Conformal Conic (LCC) projection with modified Everest as spheroid.

Base Map Preparation

The base map is referred to a map containing the basic reference details of infrastructure information such as road, railway, water body, etc., which act as points of reference while transferring the thematic details from topographical map on 1:50,000 scale. The base map details are further updated using satellite image. The various thematic layers of the study area is shown in the image Plate 15.1.

Ground Truth Verification

The ground truth verification is done for land use/land cover analysis for the doubtful areas. Spatial locations of all land use/land cover are taken using Garmin GPS instrument. Ground truth verification of doubtful areas enhances the interpretation accuracy. The doubtful areas were physically verified in the field and field observation about the terrain conditions, and land use patterns were noted for use in the modification of thematic details.

Supervised Classification

In supervised classification, the analyst makes use of prior knowledge available from topographic, photo-interpreted maps and field surveys while training the computer by way of assigning "training areas" about different features that are to be classified by the computer. Signature is grouped and assigned to respective classes.

Various supervised classification methods are used to assign an unknown pixel to one of a number of classes. Among them, the most frequently used classification algorithms are: Minimum distance to mean; Parallelepiped classification and Maximum Likelihood algorithm

Normalized difference vegetation index (NDVI) is another parameter indicating status of biomass in the area. NDVI has been calculated using the following formula:

$$\text{NDVI} = \frac{\text{NIR} - \text{Red}}{\text{NIR} + \text{Red}}$$

Land Use/Land Cover Change Detection

Statistics of land use changes are used in GIS environment. For change detection, the data of Landsat (TM) for the year 2001 and IRS LISS-III for the year 2003 were used. Classification of individual data sets and pixel-to-pixel comparison through intersection Matrix analysis has been carried out to assess the change in land use/land cover between the years 2001 and 2003.

The total number of classes is varying from area to area. The imagery has to be interpreted and ground checked for corrections.

Result and Discussion

Remote sensing in combination with GIS is very useful for the preparation of land use/land cover maps of an area. A digital classification technique has been carried out for the part of Udaipur City (South) area. IRS LISS-III and Landsat (TM) data set has been used and supervised classification techniques have been carried out for digital classification.

The change detection studies have been carried out for the Udaipur City (South) to determine the change in land used/land cover over a period of time. The change has been assessed between the years 2001 and 2003. The result obtained for the land use/cover analysis and the change detection studies of the city area are presented and discussed in this study.

Land Use/Land Cover Analysis of Udaipur City (South)

The land use/land cover categories such as settlements, agriculture natural vegetation, water bodies, bare and dense scrub land have been identified using supervised classification techniques.

The major land cover of the study area is settlement. It covers an area of 3300.65 hectare. The bare land in the study is 2993.7 hectare. The smallest area is covered by the water body and it is 180.979 hectare. It is mainly of the 'Lake Pichola'.

The land use/land cover pattern of Udaipur City (South) is shown in satellite image 1 and 2 of Landsat (TM) and IRS LISS-III data (Plate 15.1 and Plate 15.2). The land cover statistics are given in Table 15.1.

Table 15.1

Land Use/Land Cover Statistics for Udaipur City (South)
(As per IRS Satellite data, January 2003)

Land use/land cover classes	*Area (in hectare)*
Agriculture land	633.542
Natural vegetation	174.989
Dense scrub land	564.595
Bare land	2993.70
Settlement	3300.65
Water body	180.979
Total	7848.46

The result of the study is discussed in the following paragraphs.

Agriculture Land

Agriculture area is clearly delineated through the satellite data. In the image, bright red tone and smooth texture shows the cropland. On the aggregate classified output 633.542 hectare of land in the study area is agriculture land.

Natural Vegetation

The black red colours are showing the natural vegetation in the image. The area of Gulab Bagh (Sajjan Niwas Garden) is clearly seen under the natural vegetation in the satellite imagery. The area covered by natural vegetation in the study area is 174.989 hectare.

These occupy relatively higher topography uplands with scrub. These lands are generally prone to degradation or erosion. This category appears on satellite imagery as brown to greenish blue patches of varying sizes with irregular shapes and are usually associated with uplands. The area with dense scrub is 564.595 hectare as per January 2003 satellite data.

Bare Land

This category appears on satellite imagery as light yellow to brown of varying sizes. The bare land in the study area is 2993.70 hectare. They occupy higher topography – high grounds without scrub.

Settlement

It is defined is an area of human habitation. It occupies the largest area in the considered project. The area of estimated under this category is 3300.65 hectare.

Water Body

Ponds and lakes are the important water bodies that can be seen in the study area. The area statistics for the category is 180.9 79 hectare.

Change Detection Studies

The changes occur in land use/land cover pattern of any area with time. This may be due to man-made activities on the land or due to natural causes. The study of temporal change of land use/land cover over a fixed time interval is necessary for the proper management of any land.

The present study reveals the changes in the land use/land cover pattern of Udaipur City (South) area between the years 2001 and 2003. The highlights of the change detection study map are given in succeeding paragraphs.

The changes in the land use/land cover can said to be a positive change if wastelands have been converted to usable land, such as crop land. On the other hand, the change are said to be negative if the crop land has degraded to wastelands.

The change detection has been carried out by analysis of Landsat (TM) data acquired in January 2001 and IRS LISS-III data acquired in January 2003 spanning almost two years (2001–2003). The highlights of the change detection are given in Table 15.2.

Table 15.2

Highlights of the Change Detection Study During 2001 and 2003

Change in the Land use/Land cover	*Area (in hectare)*
Area decreases (Red)	1631.58
Areas increases (Green)	4976.87
Area unchanged (Yellow)	1240.01
Total	7848.46

Plate 15.1

Land Use/Land Cover Pattern of Udaipur City (South) as per Supervised Classification (TM Satellite Data, January, 2001)

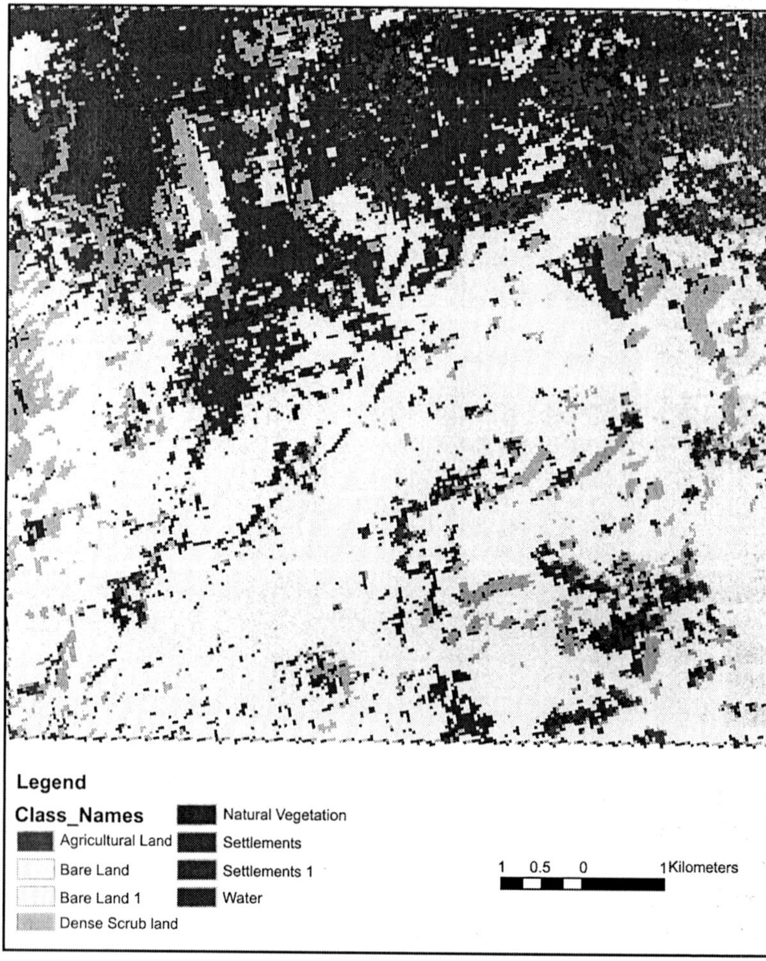

Plate 15.2

Land Use/Land Cover Pattern of Udaipur City (South) as per Supervised Classification (IRS Satellite Data, January, 2003)

The changes were observed by comparing the two supervised classification and thematic layers of the study area. The settlement's areas have been considerably increased. In 2001, the area which was bare land, had been converted into settlement in 2003. The water body more or less remains the same in the comparison period. The natural vegetation area has been decreased. The dense scrub land cover increased from 2001 to 2003. The total area depicting the increase in land use/land cover is 4976.87 hectare and the total area which shows the decrease in land use/land cover is 1631.58 hectare.

Conclusion

The quantification of various land use/land cover categories and its distribution was successfully done with the help of remote sensing and geographical information system. The remote sensing technique is found to be faster, accurate and useful for mapping or generating a spatial database for a city. The GIS tools are used in deriving desired or user-required information from the database along with statistical analysis. Because of the distinct advantages in terms of the timeliness, repetitively, real time capability of data reception and transmission and computer compatibility, remote sensing is an indispensable tool for land use/land cover mapping and trend analysis. GIS has proved to be a powerful tool for integrating these various information and generation of the geo-database.

The objectives of this study were to interpret the land use/land cover features using satellite remote sensing digital techniques and GIS. In addition, the change in the land use/land cover pattern between 2001 and 2003 has been assessed through change detection studies.

The total settlement area is 3300.65 hectare and bare land is 2993.70 hectare. The agriculture land and natural vegetation covers an area of 808.531 hectare, as per satellite data of January 2003. Comparing the land use map of 2001 and 2003, a general scenario of the major changes in land use can be arrived at. Major changes noticed are increase in the settlement at the cost of bare land. The natural vegetation has been decreased. The total area depicting the increase in land use/land cover is 4976.87 hectare from 2001 to 2003.

From the above discussion, it is concluded that by the help of remote sensing and GIS the analysis of existing land use of the area can be effectively determined and can be used for future planning purpose.

Acknowledgement

The paper is based on project work completed by Mr Ahuja supervised by Prof. P.R. Vyas, Director NRM, JRN RVU, Udaipur (as internal guide) and Mr Pradeep Bhatnagar, Superintending Geologist (RS), Department of Mines and Geology, Udaipur.

References

Anderson, J.R. (1971), 'Land Use/Land Cover Changes: A Framework for Monitoring', *Journal of Research*, Vol. 5, No. 3, U.S. Geological Survey, Government of India, pp. 143–53.

Anuta, P.E. (1977), 'Computer Assisted Analysis Techniques for Remote Sensing Data Interpretation', *Geophysics*, Vol. 42, pp. 468–81.

Bisht, B.S. and B.P. Kothyari (2001), 'Land-Cover Change Analysis of Garur Ganga Watershed Using GIS/Remote Sensing Technique', *J. Indian Soc. Remote Sensing*, Vol. 29, No. 3, pp. 165–74.

Campbell, J.B. (1983), *Mapping the Land Aerial Imagery for Land Use Information*, Association of American Geographers, Washington, DC.

Congalton, R.G. and R.A. Mead (1983), 'A Quantitative Method to Test for Consistency and Correctness in Photo Interpretation', *Photogrammetric Engineering and Remote Sensing*, Vol. 49, No. 1, pp. 67–74.

Deekshatulu, B.L. and Y.S. Rajan (1984), *Remote Sensing*, Indian Academy of Sciences, Bangalore.

Friedman, S.Z. (1980), *Mapping Urbanized Area Expansion Through Digital Image Processing of Landsat and Conventional Data*, JPL Publications, California.

Jensen, J.R. (1983), 'Urban-Suburban Land Use Analysis', Chapter 30 in R. N. Colwell (ed.), *The Manual of Remote Sensing*, Vol. 2, American Society of Photogrammetry and Remote Sensing, Falls Church, pp. 1571–966.

—— (1975), *Digital Image Processing*, Prentice-Hall, Englewood Cliffs, New Jersey.

Joseph, G. (2004), *Fundamentals of Remote Sensing*, Cambridge University Press, Cambridge, UK, pp. 1–29.

IRS-ID Handbook **(year?)**, National Remote Sensing Agency.

Lillesand, T.M., R.W. Keifer and J.W. Chipman (2004), *Remote Sensing and Image Interpretation*, John Wiley & Sons, New York.

Lindgren, David T. (1985), *Land Use Planning and Remote Sensing*, Martinus Nijhoff Publishers, Boston, pp. 101–15.

Rao, D.P. (1995), *Remote Sensing for Earth Resources*, Association of Exploration Geophysicists, Osmania University, Hyderabad, pp. 153–69.

Reddy, M. Angi (2001), *Remote Sensing and Geographic Information System*, B.S. Publications, Hyderabad.

Sokhi, B.S. (1993), 'Remote Sensing in the Study of Compatibility of Land Uses and a High Capacity Intra Urban Transportation System, New Delhi', *Asian Pacific Remote Sensing Journal*, Vol. 6.

RRSSC (2004), *Basics of Satellite Remote Sensing and Geographic Information System: Principles and Concepts*, Regional Remote Sensing Service Center, Nagpur.

Sabins, J.R. (1997), *Remote Sensing Principles and Interpretation*, 3rd Edition, W.H. Freeman & Co., New York, pp. 357–75.

Uttarwar, P.S., N.D. Sharma and B.S. Sokhi (1989), 'Satellite Remote Sensing in Urban Sprawl Mapping and Monitoring: A Case Study of Delhi', *Journal of the Indian Society of Remote Sensing*, Vol. 17, No. 3.

Vink, A.P.A. (1975), *Land Use in Advancing Agriculture*, Springer-Verlag, Berlin, p. 394.

Young, A. (1998), *Land Resources: Now and for the Future*, Cambridge University Press, Cambridge, UK, p. 319.

Glossary

absolute location: A location in geographic space given with respect to a known origin and standard measurement system, such as a coordinate system.

accuracy: The validity of data measured with respect to an independent source of higher reliability and precision.

active data: Data that can be reconfigured and recomputed in place. Spreadsheet term for data for attributes or records created by formulas within a spreadsheet.

analog: A representation where a feature or object is represented by another tangible medium. For example, a section of the earth can be represented in analog by a paper map, or atoms can be represented by ping-pong balls.

analysis: The stage in science when measurements are sorted, tested, and examined visually for patterns and predictability.

arc: (1) A line represented as a set of sequential points. (2) A line that begins and ends at a topologically significant location, represented as a set of sequential points.

arc/node: An early name for the vector GIS data structure.

area: A two-dimensional (area) feature represented by a line that closes on itself to form a boundary.

area feature: A geographic feature recorded on a map as a sequence of locations or lines that taken together trace out an enclosed area or ring that represents the feature. Example: state boundary.

area qualitative map: A type of map that shows the existence of a geographic class within areas on the map. Colours, patterns, and shades are generally used. Examples are geological maps, soils, and land-use maps.

array: A physical data structure for grids. Arrays are part of most computer programming languages and can be used for storing and manipulating raster data.

attribute: A characteristic of a feature that contains a measurement or value for the feature. Attributes can be labels, categories, or numbers; they can be dates, standardized values, or field or other measurements. An item for which data are collected and organized. A column in a table or data file.

Auto-cad: A leading CAD program by Autodesk, often interfaced with G1S packages and used for digitizing, especially floor plans and engineering graphics.

AVHRK (advanced very high resolution radiometer): An instrument on the NOAA orbiting a polar satellite that returns 1 and 4 km resolution data about the earth in four wavelengths. Used extensively for large area land-cover and vegetation mapping and weather prediction.

azimuthal: A map projection in which the globe is projected directly on a flat surface. Only one "side" of the globe can be shown at a time.

background image: A satellite image or air photo that serves as a backdrop for display and, registration purposes only rather than as a layer for analysis with the GIS.

bell curve: A common term for the normal distribution.

big eight: The eight most popular GIS packages, established by the numbers of users, particularly among people getting started newly with GIS.

block face: One side of a street on one block that is between two street intersections.

browse: A method of search involving repeated examination of records until a suitable one is found.

byte: Eight consecutive bits.

CAD (Computer aided design): Computer software used in producing technical and design-type drawings.

CALFORM: An early computer mapping package for thematic mapping.

CAM (Computer-Assisted mapping): A map projection and outline plotting program for mainframe computers dating from the 1960s.

cartographic convention: The accepted cartographic practice. For example, water is usually cyan or light blue on a world map.

cartographic elements: The primitive component part out of which a map is assembled, such as the neat line, legend, scale, titles, figure, and so on.

cartographic spaghetti: A loose data structure for vector data, with order as the only identifying property of the features.

cartography: (1) The science that deals with the principles, construction, and use of maps. (2) The science, art, and technology of making, using, and studying maps.

CD-ROM: Compact disk read-only memory, a hardware storage device capable of making extensive data and software available for distribution on removable CDs or as off-line storage for a microcomputer.

choropleth map: A map that shows numerical data (but not simply "counts") for a group of regions by (1) classifying the data into classes and (2) shading each class on the map.

colour balance: The achievement of visual harmony between colors on a map, primarily by avoiding colours that show simultaneous contrast when adjacent to each other.

colour table: The part of the header record in a digital image file that stores specifications of colours based on simple index values, which are then stored in the data part of the image file.

compression: Any technique that reduces the physical file size of data in a spatial or other data format.

compromise: A map projection that is neither area preserving nor shape preserving. An example is the Robinson projection.

compute: A data management command that uses the numerical values of one or more attributes to calculate the value of a new attribute created by the command.

computer mapping: Producing maps using the computer as the primary or only tool.

computer memory: A sequence of non-random bytes that are non-recoverable after a computer has been turned off and on again.

conic: A type of map projection involving projecting part of the earth onto a cone-shaped surface that is then cut and unrolled to make it flat.

contour interval: The vertical difference in measurement units such as metres or feet between successive contour lines on a contour map.

contour map: An isoline map of topographic elevations.

cookie-cut: A spatial operation to exclude area outside a specific zone of interest. For example, a state outline map can be used to cut out pixels from a satellite image.

coordinate pair: An easting and northing in any coordinate system, absolute or relative. Together these two values, usually termed (x, y) describe a location in two-dimensional geographic space.

CORONA: Originally a secret satellite-based remote sensing system carrying the early generation Keyhole cameras for the period 1962–74. Very high resolution monochrome data from Corona, some covering the United States are now available on-line through the USGS's EROS data centre.

credits: A cartographic element in which the sources, authorship, and ownership of the map and the map attributes are cited, often including a date or reference.

data: A set of measurements or other values, such as text for at least one attribute and at least one record.

data dictionary: (1) A catalogue of all the attributes for a data set, along with all the constraints placed on the attribute values during the data definition phase. Can include the range and type of values, category lists, legal and missing values, and the legal width of the field. (2) The part of a database containing information about the files, records, and attributes rather than, just the data.

data exchange: The exchange of data between similar GIS packages but groups with a common interest.

data exchange format: The specific physical data format in which exchange of data between similar GIS packages takes place.

data extremes: The highest and lowest values of an attribute, found by selecting the first and last records after sorting.

data format: A specification of a physical data structure for a feature or record.

data mining: Revisiting existing data to explore for new relationships using new and more powerful tools for analysis and display.

data model: A logical means of organization of data for use in an information system.

data retrieval: The ability of a database management system to get back from computer memory records stored there previously.

data structure: The logical and physical means by which a map feature or an attribute is digitally encoded.

data transfer: The exchange of data between non-communicating computer systems and different GIS software packages.

database: (1) A collection of data organized in a systematic way to provide access on demand. (2) Any collection of data accessible by computer. (3) The body of data that can be used in a database management system. A GIS has both a map and an attribute database.

database manager: A computer program or set of programs that allows a user to define the structure and organization of a database, to enter and maintain records in the database, to perform sorting, data reorganization, and searching, and to generate useful products such as reports and graphs.

datum: A base reference level for the third dimension of elevation for the earth's surface. A datum can depend on the ellipsoid, the earth model, and the definition of sea level.

dependent variable: The variable on the left of the equals sign in a formula model, whose values are determined by the values of the other variables and constants.

design loop: The iterative process in which a GIS map is created, examined for design, improved, and then replotted from the modified map definition until the user is satisfied that a good design has been reached.

device independence: The ability of software to run with little difference from a user's perspective on any computer or on any specialized device, such as a printer or plotter.

digital elevation model: A data format for digital topography, containing an array of terrain elevation measurements.

digital orthophoto quad (DOQ): One element of a national mapping effort to cover the lower 48 United States at a 1-metre ground resolution with monochrome air photos in digital format with a 1:12,000 equivalent ground extent. Collections of DOQs are distributed compressed on CD-ROM.

digitizing: Also called semi-automated digitizing. The process in which geocoding takes place manually; a map is placed on a flat tablet and a person traces out the map features using a cursor. The locations of features on the map are sent back to the computer every time the operator of the digitizing tablet presses a button.

digitizing tablet: A device for geocoding by semi-automated digitizing. A digitizing tablet looks like a drafting table but is sensitized so that as a map is traced with a cursor on the tablet, the locations are identified, converted to numbers, and sent to the computer.

dimensionality: The property of geographic features by which they are capable of being broken down into elements made up of points, lines, and areas. This corresponds to features being zero-, one-, and two-dimensional. A drill hole is a point, a stream is a line, and a forest is an area, for example.

disease transmission: Passing a disease from one individual to another.

distortion: The space distortion of a map projection, consisting of warping of direction, area, and scale across the extent of the map.

distributed network: A network-connected set of locations each storing one element of a system. A distributed GIS may have the GIS software, running on a workstation but use data dispersed at many computer storage locations over a local or wide area network.

DLG: A vector format used by the USGS for encoding the line on large-scale digital maps.

dot map: A map type that uses a dot symbol to show the presence of a feature, relying on a visual scatter to show spatial pattern. Most often used where point features are the GIS data, but dots can be scattered at random throughout areas.

download: To move a file across a network for eventual residence locally.

dropout: The loss of data due to scanning at coarser resolution than the map features to be captured. Features smaller than half the size of a pixel can disappear entirely.

edge matching: The GIS or digital map equivalent of matching paper maps along their edges. Features that continue over the edge must be "zipped" together and the edge dissolved. To edge-match, maps must be on the same projection, datum, ellipsoid, and scale and show features captured at the same equivalent scale.

elevation: The vertical height above a datum, in units such as metres or feet.

empowerment: Placing power in the hands of the citizen by providing effective and timely information.

end node: The last point in an arc that connects to another arc.

environmental assessment: The collective consideration of the local environment and its threats to individuals by measurement and analysis.

EOSDIS: The information distribution, dissemination and storage section of the EOS program, which eventually will provide fall network access to the data collected by all the various instruments and programs.

equatorial radius: The distance from the geometric centre of the earth to the surface, usually averaged to a single value for a sphere.

equirectangular: A map projection that maps angles directly to eastings and northings. A cylindrical projection, made secant by scaling the height-to-width ratio. The non-secant or equatorial version is called the Plate Carree. Credited to Marinus of Tyre, about AD 100.

equivalent: A type of map projection that preserves the area of features on maps. On an equal-area projection, a small circle on the map would have the same area as on a globe with the same representative fraction.

ESRI: ESRI is an international supplier of GIS softwares. The company was founded as an Environmental Systems Research Institute (ESRI), has its headquaters in Redlands, California, USA.

fat line: Raster representation of a line that is more than one pixel wide.

feature: A single entity that composes part of a landscape.

field variable: a geographic value that is continuous over space.

file: (1) A collection of bytes stored on a computer's storage device. (2) Data logically stored together at one location on the storage mechanism of a computer.

file header: The first part of a file, which contains metadata rather than data.

file server: A computer whose primary function is to store data and make them available on a network as part of a distributed system.

find: A database management operation intended to locate a single record or a set of records or features based on the values of their attributes.

flat file: A simple model for the organization of numbers. The numbers are organized into a table, with values for variables as entries, records as rows, and attributes as columns.

flow map: A linear network map that shows, usually by proportionally varying the width of the lines in the network, the amount of traffic or flow within the network.

FORTRAN: An early computer programming language, initially for converting mathematical formulas into computer instructions.

fourth dimension: A common way of referring to time: the first three dimensions determine location in space, the fourth dimension determines creation, duration, and destruction in time.

freeware: Software and data made available on the networks to any user at zero cost. Immense amounts of freeware, including entire operating systems and GIS packages, are available on the Internet.

FTP (file transfer protocol): A standardized way to move files between computers. It is a packet switching technique, so that errors in transmission are detected and corrected. FTP allows files, even large ones, to be moved between computers on the Internet, or another compatible network.

functional capability: One of the distinctive processes that a GIS is able to perform as a separate operation or as part of another operation.

general-purpose map: A map designed primarily for reference and navigation use.

geocode: A location in geographic space converted into computer-readable form. This usually means making a digital record of the point's coordinates.

geodesy: The science of measuring the size and shape of the earth and its gravitational and magnetic fields.

geographic coordinates: The latitude and longitude coordinate system.

geographic information science: (1) Research on the generic issues that surround the use of: GIS technology, impede its implementation, or emerge from an understanding of its capabilities. (2) The scientific use and study of methods and tools for the capture, storage, distribution, analysis, display and exploitation of geocoded information.

geographic pattern: A spatial distribution explainable as a repetitive distribution.

geographic property: A characteristic of a feature on earth, usually describable from a map of the feature, such as location, area. shape, distribution, orientation, adjacency, and so on.

geographic search: A find operation in a GIS that uses spatial characteristics as its basis. Geographic(al) information system: (1) A set of computer tools for analysing spatial data. (2) A special case of an information system designed for spatial data. (3) An approach to the scientific analysis and use of spatial data. (4) A multibillion dollar industry and business.

geographical surface: The spatial distribution traced out by a continuously measurable geographical phenomenon, as depicted or, by a map.

geography: (1) A field of study based on understanding the phenomena capable of being described and analysed with a GIS. (2) The underlying geometry and properties of the earth's features as represented in a GIS. (3) The science concerned with all aspects of the earth's surface, including natural and human divisions, the distribution and differentiation of regions, and the role of humankind in changing the face of the earth.

geometric test: A test to establish the spatial relationship between features. For example, a point feature can be given a point-in-polygon test to find if it is "contained" by an area.

GIS/LIS: A US national conference on geographic information and land information systems, sponsored by most GIS professional organizations and held annually.

GPS (Global Positioning System): An operational, system of satellites in orbits that allow their use by a receiver to decode time signals and convert the signals from several satellites to a position on the earth's surface.

graduated symbol map: A map type that varies the size of a common geometric symbol to show the amount of an attribute at points or at centroids of areas. For examples, cities could be shown with circles of area proportional to population, or census tracts could have a proportional circle divided as a pie chart at a representative point inside the tract.

GRASS: A GIS software package.

GRID: (1) A United Nations program to assemble, use, and disseminate data sets of global extent, of use to United Nations and other agencies. (2) The raster module of the ESRI Arc/Info GIS software.

ground: The part of the body of the map that is riot featured in the figure. This area can include neighbouring areas, oceans, and so on. The ground should fall lower than the figure in the visual hierarchy.

GRS80 (Geodetic Reference System of 1480): Adopted by the International Union of Geodesy and Geophysics in 1979 as a standard set of measurements for the earth's size and shape. The length of the semi-major axis is 6,378,137 metres. Flattening is 1/298.257.

GUI (graphical user interface): The set of visual and mechanical tools through which a user interacts with a computer, usually consisting of windows, menus, icons and pointers.

helpline: A telephone service available to software users for help from an expert.

hexadecimal: A counting system based on 16.

hierarchical data model: An attribute data model based on sets of fully enclosed subsets and many layers.

histogram: A graphic depiction of a sample of values for an attribute, shown as bars raised to the height of the frequency of records for each class or group of value within the attribute.

HSI: A system of specifying colour as three values: hue, saturation, and intensity, respectively.

hue: A colour as defined by the wavelength of the light reflected or emitted from the map surface.

hypothesis: A supposition about data expressed in a manner to make it subject to statistical test.

hypsometric map: A map of topography involving a colour sequence filling the spaces between successive contours, usually varying from green through yellow to brown.

image depth: The numbers of bits stored for each pixel in a digital image.

import: The capability of a GIS to bring data in an external file and in a non-native format for use within the GIS.

information system: A system designed to allow the user to be delivered the answer to a query from a database.

installed base: The number of existing implemented systems.

Internet: A network of many computer networks. Any computer connected to the Internet can share any of the computers accessible through the network. The Internet shares a common mechanism for communication, called a protocol. Searches for data, tools for browsing, and so on are available to ease the tasks of "surfing" the Internet.

interval: Data measured on a relative scale but with numerical values based on an arbitrary origin. Examples are elevations based on mean sea level or coordinates.

isoline map: A map containing continuous lines joining all points of identical equal value.

key attribute: A unique identifier for related records that can serve as a common thread throughout the files in a relational database.

label placement rules: The set of rules that cartographers use, when adding map text, place names, and labels to features. Some rules are generic to the map as a whole, while others relate to point, line, and area features specifically. Well designed maps follow the label placement rules and use them to resolve conflicts between the labels, as labels should never be plotted over each other **label point:** A point digitized within a polygon and assigned its label or identifier for use in topological reconstruction of the polygon.

land-use map: A map showing the human use to which land is put at a given time. Categories could be pasture, forest land, agricultural land, wasteland, and so on.

Landsat: A US government satellite program collecting data about the earth's surface in the visible and infrared parts of the spectrum. Two instruments, the multi-spectral scanner (79-metre resolution) and the thematic mapper (30-metre resolution), have been used. Landsat 7 is the next to be launched, for which the data will return to the public domain.

landscape: The part of geographic space showable on a map, including all its features.

latitude: The angle made between the equator, the earth's geometric centre, and a point on or above the surface. The south pole has latitude -90 degrees, the north +90 degrees.

layer: A set of digital map features collectively (points, lines, and areas) with a common theme in coregistration with other layers. A feature of GIS and most CAD packages.

learning curve: The relationship between learning and time. A steep learning curve means that much is learned quickly (usually thought to be the opposite). A difficult learning curve is one where learning takes place slowly, over a long period.

legend: The map element that allows the map user to translate graphic map symbols into ideas, usually by the use of text.

line: A one-dimensional (length) map feature represented by a string of connected coordinates.

line feature: A geographic feature recorded on a map as a sequence of locations tracing out a line. A stream is an example.

linear relationship: A straight-line relationship between two variables such that the value of the dependent variable is a gradient times the independent variable plus a constant.

local area network: An arrangement of computers into a cluster, with network linkages between computers but no external link. Usually, this allows sharing data and software licenses, or the use of a file server.

location: A position on the earth's surface or in geographic space definable by coordinates or some other referencing system, such as a street address or space indexing system.

longitude: The angle formed between a position on or above the earth, the earth's geometric centre, and the meridian passing through the centre of the observing instrument in Greenwich, England, as projected down onto the plane of the earth's equator or viewed from above the pole. Longitudes range from -180 (180 degrees West) to +180 (180 degrees East).

map design: The set of choices relating to how a map's elements are laid out, how symbols such as colours are selected, and how the map is produced as a finished tangible product. The process of applying cartographic knowledge and experience to improve the effectiveness of a map.

map overlay: Placing multiple thematic maps in precise registration, with the same scale, projections, and extent, so that a compound view is possible.

map title: The text that identifies the coverage and content of a map. This is usually a major map element, and can be worded to show the map theme, or the map's content.

mask: A map layer intended to eliminate or exclude areas not needed for mapping and analysis.

mean sea level: A local datum based on repeated measurements of sea level throughout all of its normal cycles, such as tides and seasonal change. The basis for elevations on a map.

menu: A component of a user interface that allows the user to make selections and choices from a preset list.

metadata: Data about data. Index-type information pertaining to the entire data set rather than the objects within the data set. Metadata usually includes the date, source, map projection, scale, resolution, accuracy, and reliability of the information, as well as data about the format and structure of the data set.

metaphor: For a GUI, the physical analogy for the elements with which the user will interact. Many computer GUTS use the desktop as a metaphor, with the elements of a calendar, clock, files and file cabinets, and so on.

metric system: A system of weights and measures accepted as an international standard as the Systeme International d'Unites (SI) in 1960. The metre (meter in the United States) is the unit of length.

military grid: A coordinate system based on the transverse Mercator projection, adopted by the US Army in 1940s and used extensively for world mapping.

missing value: A value that is excluded from arithmetic calculations for an attribute because it is missing, not applicable, or is corrupted, and has been signified as such.

modelling: The stage in science when a phenomenon under test is sufficiently well understood that an abstract system can be built to simulate the real system.

mosaic: The GIS or digital map equivalent of matching paper maps along their edges. Features that continue over the edge must be "zipped" together and the edge dissolved. To edge match, maps must be on the same projection, datum, ellipsoid, and scale, and show features captured at the same equivalent scale. See also edge matching.

multitask: The ability of a computer's operating system or GIS to handle more than one process at once: for example, editing and running a command sequence while extracting data from the database and displaying a map.

National Spatial Data Clearinghouse: A World Wide Web resource that serves as a cross-reference point for the distributed database of all US government public-domain and other geographic information.

network conference group (also network news group, usenet group, List): An Internet resource on which users with similar interests can share broadcast exchanges on information. Several major GIS groups exist, including comp.infosystems.gis; also called GIS-L.

NOAA (National Oceanic and Atmospheric Administration): A part of the US Department of Commerce, that is the provider of digital and other maps for navigation, weather prediction, and for physical features of the United States.

node: (1) At first, any significant point in a map data structure. Later, only those points with topological significance, such as the ends of lines. (2) The end of an arc. At first, any significant point in a map data structure. Later, only those points with topological significance, such as the ends of lines.

node snap: Instructing the GIS software to make multiple nodes or points in a single node so that the features connected to the nodes match precisely, say at a boundary.

nominal: A level of measurement at which only subjective information is available about a feature. For a point, for example, the name of the place.

normal distribution: A distribution of values symmetrically about a mean with a given variance.

null hypothesis: The state opposite to that suggested in a hypothesis, postulated in the hope of rejecting its form and therefore proving the hypothesis.

oblique: A map projection in which the centre line of the map is not at right angles to the earth's geographic coordinates, following neither a single parallel nor a meridian.

observation: The process of recording an objective measurement.

Open/GIS: An active effort to assure interoperability among GIS software packages by specifying a standard set of functions and a common user interface.

operating system: The suite of software programs and utilities necessary for the control and use of a computer, including as a minimum the management of files and use of the computer's processor.

orthophoto map: An image map that is an air photo, corrected for topographic and other effects. A specific type of mapping program, at 1:12,000, by the USGS.

parallel: A line of constant latitude. Parallels get shorter toward the poles, becoming a point at the pole itself.

PC (Personal Computer): A self-contained microcomputer, providing the necessary components for computing, including hardware, software, and a user interface.

physical structure: The mechanical mapping of a section of computer memory onto a set of files or storage devices.

place name: A text cartographic element that links a place name to a feature by placing it close to the symbol to which it corresponds, such as a city name as text next to a filled circle.

point: A zero-dimensional map feature, such as a single elevation mark as specified by, at least two coordinates.

population: The total body of objects from which a sample is taken for measurement.

PostScript: Adobe Corporation's page definition language. An interpreted language for page layout designed for printers but used as an industry standard for vector-graphics.

precision: The number of digits used to record a measurement or which a measuring device is capable of providing.

prediction (1) The scientific ability to forecast the outcome of a process in advance. (2) The ability of a model to provide information beyond that for which measurements are available.

prime meridian: The line traced out by longitude zero and passing through Greenwich, England. The prime meridian forms the origin for the longitude part of the geographic coordinates and divides the eastern and western hemispheres.

proceedings: The formal record of the papers and other prepared presentations at a conference. Usually available to conference attendees, then distributed as a softcover book.

professional publication: Books, journals, and other information designed primarily for those using GIS technology as part of their job.

proprietary format: A data format whose specification is a copyrighted property rather than public knowledge.

quad tree: A way of compressing raster data based on eliminating redundancy for attributes within quadrants of a grid.

RAM: The part of a computer's memory designed for rapid access and computation.

raster: A data structure for maps based on grid cells.

ratio: A level of measurement at which numerical information is available about a feature, based on an absolute origin. For land parcels, for example, the assessed value in dollars would be an example, the value zero having real meaning.

real map: A map that has been designed and plotted onto a permanent medium such as paper or film. It has a tangible form and is a result of all the design and compilation decisions made in constructing the map, such as choosing the scale, setting the legend, choosing the colours, and so on.

listic perspective map: A map of a three-dimensional surface showing a coloured or shaded image draped over a topographic surface and viewed in perspective. Record: A set of values for all attributes in a database. Equivalent to the row of a data table.

I-Aerence map: A highly generalized map type designed to. show general spatial properties of features. Examples are world maps, road maps, atlas maps, and sketch maps. Sometimes used in navigation, often with a limited set of symbols and selected data. A cartographic base reference map is often the base layer in a GIS.

relate: A DBMS operation that merges databases through their key attributes to restructure them according to the needs of a user's query, rather than as they are stored physically.

relational DBMS: A database management system based an the relational data model. Relational model: A data model based on multiple flat files for records, with dissimilar attribute structures, connected by a common key attribute.

relative location: A position described solely with reference to another location.

remote login: The ability of a computer user to log directly into another computer through a network connection or a modem.

remotely sensed data: Data collected by a sensor that is not in direct contact with the area being mapped. Active remote sensing involves transmitting a beam that is detected after reflection; passive remote sensing simply measures light from the sun being reflected by objects being sensed. Similar instruments for remote sensing can operate from aircraft or satellites.

renumbering: Use of the DBMS to change the ordering or ranges of attributes. Also, especially in raster GISs, to change the numbers within grid cells into categories.

report: A listing of all the values of attributes for all records in a database. A report is often printed as 'a table, for 'verification against source material, and for validation by examination.

report generator: The part of a database management system that can produce a listing of all the values of attributes for all records in a database.

representative fraction: The ratio of a distance as represented on a map to the equivalent distance measured on the ground. Typical representative fractions are 1:1 million, 1:100,000, and 1:50,000.

residual: The amount left when the observed value of the dependent variable has subtracted from it that predicted by a model, in units of the dependent variable. Resource management: The intentional control or influence of environmental elements to accomplish particular goals.

restrict: Part of the query language of a DBMS that allows a subset of attributes to be selected out of the flat file.

retrieval: The ability of a database management system or GTS to get back from computer memory records that were stored there previously.

RGB: The system of specifying colours by their red, green, and blue saturations ring: A line that closes upon itself to define an area.

rubber sheeting: A statistical distortion of two map layers so that spatial co-registration is accomplished, usually at a set of common points.

run-length encoding: A way of compressing raster databased on eliminating redundancy for attributes along rows of a grid.

sample: A subset of a population selected for measurement.

saturation: The amount of colour applied per unit area. Perceptually, saturated colours appear rich or solid, whereas low saturation colours look washed out or pastel-like.

scale: (1) The geographic property of being reduced by a representative fraction. Scale is usually depicted on a map or can be calculated from features of known size. (2) The part of the map display that shows the scale of the map figure as either an expression of values (the representative fraction as a number) or as a graphic, usually a line on the map labelled with an equivalent and whole number length on the ground, such as 1 kilometre or 1 mile.

scaleless: The characteristic of digital map data in abstract form of being usable and displayable at any scale, regardless of the scale of the map used to geocode the data.

scanning: A form of geocoding in which maps are placed on a surface and scanned by a light beam. Reflected light from every small dot or pixel on the surface is recorded and saved as a grid of digits. Scanners can work in black and white, in gray tones, or in colour.

scientific approach: A method for rationally explaining observations about the natural/and human world.

scientific visualization: Use of the human visual processing system assisted by computer graphics, as a means of direct analysis and interpretation of information.

search: Any database query that results in successful retrieval of records.

search engine: A software tool designed to search the Internet and WWW for documents meeting the users query. Examples: Yahoo, Alta Vista, and Hotbat. Secant: A map projection in which the surface used for the map "cuts" the globe at the map's representative fraction. Along this line there is distortion-free mapping of the geographic space. Multiple cuts are possible, for example, on a conic projection.

select: A DBMS command designed to extract a subset of the records in a database.

server: A computer connected to a network whose primary function is to act as a library of information that other users can share.

shareware: Data or software placed in the public domain for distribution, but whose use or support involves the payment of a' (usually token) fee to the author.

sift: To eliminate features that are smaller than a minimum feature size.

simulated hill-shaded map: A map in which an apparent shading effect of raised topography is produced by computer (or manually) so that the land surface appears differentially illuminated, as it would in low sun angles naturally. Simultaneous contrast: The tendency for colours at the apposite ends of the primary scale to "jump" when placed together: for example., red and green.

SLF: An early Defense Mapping Agency data format.

sliver: A very small, narrow polygon caused by data capture or overlay error that does.

slope: The constant of multiplication in a linear relationship, that is, the rate of increase of a straight line up or down.

snap: Forcing two or more points within a given radius of each other to be the same point, often by averaging their coordinates.

software package: A computer program application.

sort: To place the records within an attribute in sequence according to their value.

spatial data: Data that can be linked to locations in geographic space, usually via features on a map.

spatial distribution: The locations of features or measurements observed in geographic space.

SPOT (System Proprietaire pour (Observation de la Terra): A French remote sensing satellite system with 10-and 20-metre resolution and stereo capability. Spreadsheet: A computer program that allows the user to enter numbers and test into a table with rows and columns and then maintain and manipulate those numbers using the table structure.

SQL (Structured Query Language): A standard language interface to relational database management systems.

standard deviation: A normalized measure of the amount of deviation from the mean within a set of values. The mean deviation from the mean.

standard distance: A two-dimensional equivalent of the standard deviation, a normalized distance built from the standard deviations of the eastings and northings for a set of points.

standard parallel: A parallel on a map projection that is secant and therefore distortion-free.

state plane: A coordinate system common in utility and surveying applications in the lower United States and based on zones drawn state by state on transverse Mercator and Lambert conformal conic projections.

stepped statistical surface: A map type in which the outlines of areas are "raised" to a height proportional to a numerical value and viewed in apparent perspective. The areas then appear as columns, with a column height proportional to value.

stream mode: A method of encoding in semi-automated digitizing, in which a continuous stream of points follows a press of the cursor button. This mode is often used for digitizing long features such as streams and coastlines. It can generate many data very quickly, so is often weeded by generalization.

SURFACE II: An early computer mapping package from the Kansas Geological Survey.

SYMAP: An early multipurpose computer mapping package.

symbol: An abstract graphic representation of a geographic feature for representation on a map. For example, the feature could be a canal, the symbol a blue line of a given thickness.

table: (1) An arrangement of attributes and records into rows and columns to assist display and analysis. (2) Any kind of organization by placement of records into rows and columns.

TCP/IP: A network communications protocol that forms the basis of most of the communications on the Internet.

temporary map: A map designed for use as an intermediate product in the GIS process, and not usually subjected to the normal map design sequence.

thematic map: A map designed primarily to show a "theme," a single spatial distribution or pattern, using a specific map type.

tick: A blood-sucking arachnid insect of the family *Izodoirlecz*. TIF: An industry-standard raster graphic or image format. TIGER: A map data format based on zero, one, and two cells, used by the US Census Bureau in street level mapping of the United States.

TIN: A vector topological data structure designed to store the attributes of volumes, usually geographic surfaces.

tolerance: The distance within which features are assumed to be erroneously located in different versions of the same thing.

topographic map: A map type showing a limited set of features but including the minimum information about elevations or landforms. Example: contour maps. Topographic maps are common for navigation and for use as reference maps.

topologically clean: The status of a digital vector map when all arcs that should be connected are connected at nodes with identical coordinates, and the polygons formed by connected arcs have no duplicate, disconnected or missing arcs.

topology: (1) The property that describes adjacency and connectivity of features. A topological data structure encodes topology with the geocoded features. (2) The numerical description of the relationships between geographic features, as encoded by adjacency, linkage, inclusion or proximity. Thus, a paint can be inside a region, a line can connect to others, and a region can have neighbours. The numbers describing topology can be stored as attributes in the GIS and used for validation and other stages of description and analysis.

toxic release: Release of a toxic substance into the environment, such as the air venting of a poisonous gas.

transparent overlay: An analog method for map overlay, where maps are traced or photographed onto transparent paper or film and then overlain mechanically.

transverse: A map projection in which the axis of the map is aligned from pole to pole rather than along the equator.

units: The standardized measurement increments for values within an attribute.

unix: A computer operating system that has been made workable on virtually every possible computer and has become the operating system of choice for workstations and science and engineering applications.

unsupervised classification: The grouping of pixels by their numerical spectral characteristics without the intervention of direct human guidance.

update: Any replacement of all or part of a data set with new or corrected data. Upward compatibility: The ability of software to move on to a new version with complete support for the data, scripts. functions, and so on, of earlier versions.

Census Bureau: A part of the Department of Commerce that provides maps in support of the decennial (every 10 years) census of the United States, especially the census of population.

user group: Any formal or informal organization of users of a system that share experiences, information, news, or help among themselves.

user interface: The physical means of communication between a person and a software program or operating system. At its most basic, this is the exchange of typed statements in English or a program-like set of commands.

USGS (United States Geological Survey): A part of the Department of the Interior and a major provider of digital map data for the United States.

UTM (Universal Transverse Mercator): A standardized coordinate system based on the metric system and a division of the earth into sixty, 6-degree-wide zones. Each zone is projected onto a transverse Mercator projection, and the coordinate origins are located systematically. Both civilian and military versions exist.

validation: The process by which entries placed in records in an attribute data file and the map data captured during digitizing or scanning, are checked to make sure that their values fall within the bounds expected of them aud that their distribution makes sense within the GIS.

value: The content of an attribute for a single record within a database. Values can be text, numerical, or codes.

variance: The total amount of disagreement between numbers. Variance is the sum of all values with their means subtracted and then squared.

vector: A map data structure using the point or node and the connecting segment as/the basic building block for representing geographic features.

vector (disease): A disease that requires an intermittent host for transmission, such as the mosquito in malaria.

verification: A procedure for checking the values of attributes for all records in a database against their correct values.

version: An update of software. Complete rewrites are usually assigned entirely new version numbers (version 3), while fixes and minor improvements are given decimal increments (version 3.1).

virtual map: A virtual map is one that has yet to be realized as a tangible map, it exists as a set of possible maps. For example, the same digital base map and set of numbers can be an entire series of possible virtual maps, yet only one may be chosen to be rendered as a real map on a permanent medium.

VISICALC: A spreadsheet package for first-generation microcomputers. Supported data tables in flat files.

visual centre: A location on a rectangular map, about 5 per cent of the height above the geometric centre, to which the eye is drawn perceptually.

visual hierarchy: The perceptual organization of cartographic elements such that they appear visually to lie in a set of layers of increasing importance as they approach the viewer.

volume: A three-dimensional (volume) feature represented by a set of areas that encloses part of a surface. In a GIS, usually the top surface only.

VPF (Vector product format): A data transfer standard within DIGEST for vector data.

warping: See rubber sheeting.

workstation: A computing device that includes as a minimum a microprocessor, input and output devices, a display, and hardware and software for connecting to a network. Workstations are designed to be used together on local area networks and to share data, software, and so on.

World Data Bank: One of the first digital maps of the world, published in two versions by the Central Intelligence Agency in the 1960s.

World Wide Web (WWW or W3): A distributed database of information stored on servers connected by the Internet and special-purpose software for browsing, searching and downloading.

X-Windows: A public-domain GUI built on the Unix operating system and computer graphics capabilities, written and supported by the Massachusetts Institute of Technology and the basis of most workstation shareware on the Internet. zero/one/two cell: TIGER terminology for point, line, and area, respectively. zip: See mosaic.

zone (for a coordinate system): The region over which the coordinates relate with respect to a single origin. Usually, some part of the earth or a state.

SPACE CENTRES AND UNITS IN INDIA

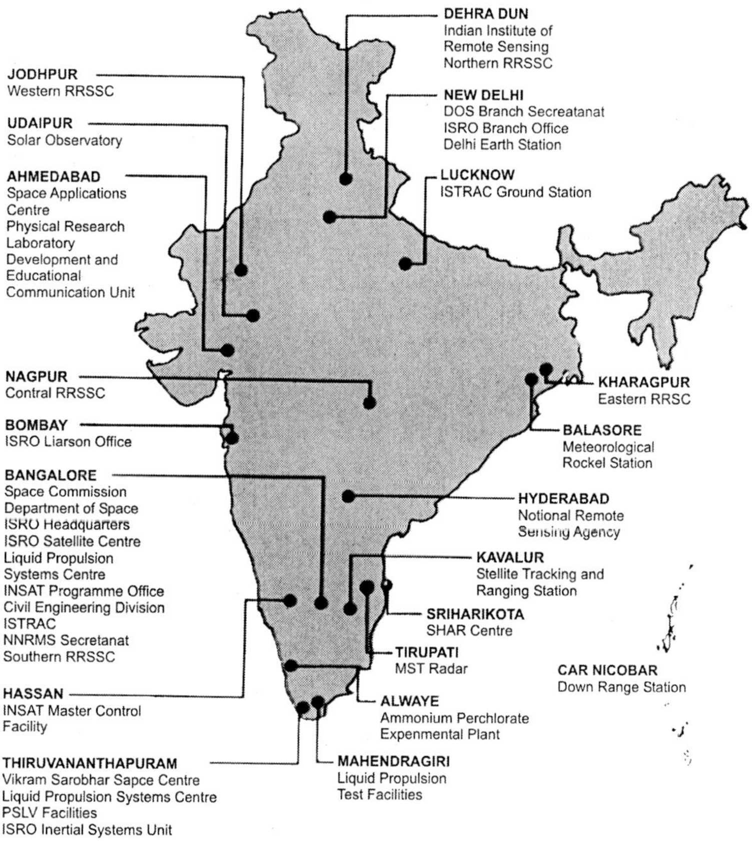

Contributors

Akshay Manohar, Project Fellow (Internship), Remote Sensing Instruments, Bagh Amberpt, Hyderabad.

Amarjeet, Scientist, Institute of Geo Information and Sustainable Development, Jodhpur.

Ashwini Kumar Vaishnav, Project Fellow (Internship), Regional Remote Sensing Service Centre, Department of Space, Government of India, CAZRI Campus, Jodhpur.

Chanchal Choudhary, Scientist, Regional Remote Sensing Service Centre (RRSSC), Jodhpur.

Deepak Kumar Ahuja, Project Fellow (Internship), Mines and Geology Department, Government of Rajasthan, Udaipur.

J.P. Sharma, Assistant Professor, Department of Geography, University of Kota, Kota.

Jagdish Soni, Project Fellow (Internship), BISAG, Gandhinagar (Gujarat).

L.R. Patel, Associate Professor, Department of Geography, JRN Rajasthan Vidyapeeth, Udaipur.

Maya Poonia, Project Fellow (Internship), B.M. Birla Science and Technology Centre, Jaipur.

Neeraj Chahar, Project Fellow (Internship), B.M. Birla Science and Technology Centre, Jaipur.

P.R. Vyas, Professor of Geography, Mohanlal Sukhadia University, Udaipur, Rajasthan.

Pankaj Rawal, Assistant Professor, Department of Geography, & NRM, JRN Rajasthan Vidyapeeth, Udaipur.

Pradeep Purohit, Department of Geography, B.N.P.G. College, Udaipur.

R.P. Naraniwal, Associate Professor, Department of Geography, JRN Rajasthan Vidyapeeth, Udaipur.

R.P. Sharma, (Sr. RAS) Scientist and Secretary, Urban Improvement Trust, Udaipur.

Rahul Tiwari, Project Fellow (Internship), Centre of Geo-informatics Application in Rural Development, National Institute of Rural Development, Hyderabad.

Reena Devi, Project Fellow (Internship), Soil Survey and Land Evaluation, Hissar, Haryana.

Shankar Bagale, Project Fellow (Internship), National Remote Sensing Agency, Department of Space, Government of India, Hyderabad.

Somendra K. Sharma, Project Fellow (Internship), B.M. Birla Science and Technology Centre, Jaipur.

Sunita Singh, Head and Associate Professor, Department of Geography, JRN Rajasthan Vidyapeeth, Udaipur.

Veena Charan, Project Fellow (Internship), RRSSC, DOS-ISRO, Nagpur.

Yuv Raj Singh Rathore, Assistant Professor, Department of Geography, & NRM, JRN Rajasthan Vidyapeeth, Udaipur.